Palgrave Studies in Digital Business & Enabling Technologies

Series Editors
Theo Lynn
Irish Institute of Digital Business
DCU Business School
Dublin, Ireland

John G. Mooney
Graziadio Business School
Pepperdine University
Malibu, CA, USA

This multi-disciplinary series will provide a comprehensive and coherent account of cloud computing, social media, mobile, big data, and other enabling technologies that are transforming how society operates and how people interact with each other. Each publication in the series will focus on a discrete but critical topic within business and computer science, covering existing research alongside cutting edge ideas. Volumes will be written by field experts on topics such as cloud migration, measuring the business value of the cloud, trust and data protection, fintech, and the Internet of Things. Each book has global reach and is relevant to faculty, researchers and students in digital business and computer science with an interest in the decisions and enabling technologies shaping society.

More information about this series at
http://www.palgrave.com/gp/series/16004

Theo Lynn • John G. Mooney
Lisa van der Werff • Grace Fox
Editors

Data Privacy and Trust in Cloud Computing

Building trust in the cloud through assurance
and accountability

Editors
Theo Lynn
Irish Institute of Digital Business
DCU Business School
Dublin, Ireland

John G. Mooney
Graziadio Business School
Pepperdine University
Malibu, CA, USA

Lisa van der Werff
Irish Institute of Digital Business
DCU Business School
Dublin, Ireland

Grace Fox
Irish Institute of Digital Business
DCU Business School
Dublin, Ireland

ISSN 2662-1282   ISSN 2662-1290   (electronic)
Palgrave Studies in Digital Business & Enabling Technologies
ISBN 978-3-030-54659-5   ISBN 978-3-030-54660-1   (eBook)
https://doi.org/10.1007/978-3-030-54660-1

This Palgrave Macmillan imprint is published by the registered company Springer Nature Switzerland AG.
The registered company address is: Gewerbestrasse 11, 6330 Cham, Switzerland

PREFACE

This is the sixth volume in *Palgrave Studies in Digital Business and Enabling Technologies*, a series that aims to contribute to multi-disciplinary research on digital business and enabling technologies in Europe. Cloud computing is the computing paradigm of choice for next generation applications and is a fundamental building block in social media and the Internet of Things, amongst others. Much of the innovation we associate with digital transformation and the socio-economic benefits of a 'Digital Society' are built on cloud computing and the exchange of data between various actors and systems in this ecosystem. Despite the massive opportunity in cloud computing, in terms of employment and GDP, Europe lags other major markets, both in cloud provision and adoption. This book focuses one of the most significant barriers to cloud computing adoption, trust and privacy.

Cloud computing is a rapidly evolving borderless technology, a product of advances in information and communications technology and globalisation. Consequently, it poses unique opportunities and challenges. Academics, industry, and policymakers alike have proposed a wide array of solutions for addressing legally certain secure and privacy-aware cloud computing while at the same time allowing firms fair use of data to remain competitive. Academic interest in and exploration of trust and privacy in the context of cloud computing has expanded rapidly over the last two decades. This book aims to provide an overview of progress in the field to date paying attention to key areas that are likely to be of interest to the cloud computing industry as well as trust and privacy experts and researchers. It is designed to be useful to students, researchers and practitioners seeking to gain access to knowledge about cloud computing and the common issues associated with it, as well as to cloud computing experts interested in discovering fresh perspectives on one of the most commonly cited barriers to adoption.

To achieve this, we invited contributors from a range of disciplines and theoretical perspectives to highlight opportunities for insight, integration and further research as the field matures. Our book brings together perspectives from psychology, law, information systems, ethics and computer science to provide a reference point for current knowledge on trust and privacy in cloud computing and to open debate on how the field can be progressed in a way that is useful to practice. The book is organised into seven chapters:

Chapter 1 introduces cloud computing, defines trust, and explores our current understanding of the range of antecedents that drive trust decisions with a discussion on how these might apply in the cloud context. This chapter also discusses some of the approaches proposed to overcome trust barriers to cloud computing, and outlines a framework for exploring assurance and accountability in the cloud.

Chapter 2 draws on the literature on trust and control to examine the contractual issues associated with cloud computing. The chapter discusses the theoretical relationships between trust, contracts and contract law, and outlines common terms and conditions in cloud computing contracts along with the issues these present. The chapter discusses how both the nature of cloud computing and contracts are evolving, and how trust can be a useful lens for developing individual vigilance and industry regulation.

Chapter 3 continues with a legal perspective considering how cloud computing can be regulated across jurisdictions with a specific focus on differences between EU and US approaches to data privacy law. The chapter explores the challenges associated with regulating borderless cloud computing and a review of recent cases in the area of data storage and location.

Chapter 4 delves further into the issue of information privacy. The chapter outlines the range of privacy related issues that can accompany cloud computing and considers how theoretical developments and empirical literature from the field of Information Systems can offer insight in the cloud context. The chapter reviews current organisational approaches for enhancing privacy perceptions in the cloud, and concludes by highlighting some research gaps that may serve as avenues for future research.

Chapter 5 introduces the concept of corporate social privacy and discusses responsibility for data privacy in the cloud. The chapter presents a conceptualisation of organisational privacy orientations that distinguishes between control- and justice-driven privacy behaviours to identity four categories of approaches. These approaches are illustrated in a framework by reference to the privacy behaviours of key players in the cloud computing market.

Chapter 6 provides an overview of ethics and ethical theory and its practical application in the context of cloud computing. In particular, the chapter explores ethical issues related to data ownership and data privacy, the provision of cloud computing services including codes of ethics and responsible practice, and the use of cloud computing by consumers.

Chapter 7 concludes the book with a computer science perspective on the issue of trustworthy cloud computing. The chapter discusses extant computer science research on trustworthy cloud computing through the lens of security and privacy, reliability, and business integrity, and discusses technical approaches to measuring and improving cloud trustworthiness.

This book was largely written before the onset of COVID-19, a health crisis that is transforming how we live, work and interact with each other. The day to day lives of organisations and individuals have been compromised in such a way that technologies such as cloud computing are critical for ensuring that economies and society persists, albeit sub-optimally. While the benefits of cloud computing have never been more evident, so are the risks. Massive rapid transition to remote working introduces new and significant security and privacy threats and concerns. COVID-19 has accelerated digital transformation in ways unforeseen a year ago. Against this backdrop, this book provides a timely reference for considering how we will assure privacy and build trust in such a hyper-connected digitally dependent world.

Dublin, Ireland   Theo Lynn
Malibu, CA   John G. Mooney
Dublin, Ireland   Lisa van der Werff
Dublin, Ireland   Grace Fox

Acknowledgement

This book was partially funded by the Irish Institute of Digital Business. Chapter 6 was funded by the Irish Research Council's Employment Based Post-graduate Scholarship programme.

CONTENTS

ABBREVIATIONS

AES	Advanced Encryption Standard
AHP	Analytical Hierarchical Process
AI	Artificial Intelligence
API	Application Programming Interface
AUP	Acceptable Use Policy
B2B	Business-to-Business
B2C	Business-to-Consumer
C2T	Cloud-to-Thing
CapEx	Capital Expenditure
CCM	Cloud Controls Matrix
CCPA	California Consumer Privacy Act
CFIP	Concern for Information Privacy
CIPO	Company Information Privacy Orientation
CPABE	Ciphertext-Policy Attribute-Based Encryption
CS	Cryptographic Server
CSA STAR	Cloud Security Alliance Security Trust Assurance and Risk
CSB	Cloud Service Brokerage
CSC	Cloud Service Customers
CSP	Cloud Service Provider
CSR	Corporate Social Responsibility
DHT	Dynamic Hash Table
DoS	Denial of Service
DPM	Data Protection Manager
DS	Digital Signature
ECJ	European Court of Justice
EDSP	European Data Protection Supervisor
EMRS	Efficient Multi-keyword Ranked Search

ENISA	European Union Agency for Network and Information Security
EU	European Union
FaaS	Function-as-a-Service
FIPPs	Fair Information Practice Principles
FTC	Federal Trade Commission
GDPR	General Data Protection Regulation
HIPAA	Health Insurance Portability and Accountability Act
HITECH	Health Information Technology for Economic and Clinical Health Act
IaaS	Infrastructure as a Service
ICO	Information Commissioner's Office
ICT	Information and Communication Technologies
IDC	International Data Corporation
IFIP	International Federation for Information Processing
IoT	Internet of Things
IP	Intellectual Property
IPC	Internet Privacy Concerns
IS	Information Systems
ISAE	International Standard on Assurance. Engagements
ISO	International Organization for Standardization
IT	Information Technology
IUIPC	Internet Users' Information Privacy Concern
KPI	Key Performance Indicators
MCS	Monte Carlo Simulation
ODR	Online Dispute Resolution
OpEx	Operational Expenditure
PaaS	Platform as a Service
PC	Privacy Concerns
PCI-DSS	Payment Card Industry Data Security Standard
PCT	Privacy Calculus Theory
QoS	Quality of Service
SA	Supervisory Authority
SaaS	Software as a Service
SeDaSC	Secure Data Sharing in Clouds
SLA	Service Level Agreement
SMI	Service Measurement Index
SOC	Service Organization Control
SSE	Searchable Symmetric Encryption
SSRF	Server-side Request Forgery
STOVE	Strict, Observable, Verifiable Data and Execution
ToS	Terms of Service
UDHR	United Nations Declaration of Human Rights (UDHR)
VM	Virtual Machine

NOTES ON CONTRIBUTORS

Olasunkanmi Matthew Alofe is a postgraduate researcher at the University of Derby where he is pursuing a PhD in cyber security. He also holds a MSc. in Cyber Security from the University of Derby. His current research encompasses cyber security, the Internet of Things, and machine learning.

Edoardo Celeste is an Assistant Professor of Law at the School of Law and Government at Dublin City University, Ireland. He specialises in law, technology and innovation. His research interests lie in the field of digital rights and constitutionalism, social media policy and regulation, privacy and data protection law, and EU digital policy. Dr Celeste is currently the Principal Investigator of the 'Digital Constitutionalism: In Search of a Content Governance Standard' project funded by Facebook Research.

Federico Fabbrini is Full Professor of European Law at the School of Law & Government and the Principal of the Brexit Institute at Dublin City University, Ireland. He is the author of "Fundamental Rights in Europe" (Oxford University Press 2014) and "Economic Governance in Europe" (Oxford University Press 2016). He is the Editor of "The Law & Politics of Brexit" (Oxford University Press 2017) and "The Law & Politics of Brexit. Volume 2: the Withdrawal Agreement" (Oxford University Press 2020), amongst others. In 2019, he was awarded the Charlemagne Prize, a research fellowship, in recognition of his work on the Future of Europe.

Kaniz Fatema is a Lecturer in Computer Science at Aston University in the United Kingdom. Previously, she worked as a Senior Lecturer at the University of Derby and as a Research Fellow at Trinity College Dublin and University College Cork, both in Ireland. She completed her PhD in Computer Science (Information Security) at the University of Kent, and holds a MSc. in Data Communications from the University of Sheffield. She has almost a decade of research experience in information and cyber security including research on access control, data protection, compliance assurance for data protection regulations, and cloud computing.

Grace Fox is an Assistant Professor of Digital Business at Dublin City University and Research Lead for the Future of Information Systems theme at the Irish Institute of Digital Business. Her research focuses on the intersection between consumers' privacy concerns and their acceptance and use of digital technologies. Her work has been published in numerous international conferences and peer-reviewed journals including Information Systems Journal, Journal of the Association for Information Science and Technology, Communications of the Association for Information Systems, and the Journal of Cloud Computing.

Theo Lynn is Full Professor of Digital Business at Dublin City University and is Associate Dean for Strategic Projects at DCU Business School. He was formerly the Principal Investigator (PI) of the Irish Centre for Cloud Computing and Commerce, an Enterprise Ireland/IDA-funded Cloud Computing Technology Centre, and Director of the Irish Institute of Digital Business. Professor Lynn specialises in the role of digital technologies in transforming business processes with a specific focus on cloud computing, social media and data science.

Valerie Lyons is the Chief Operations Officer of BH Consulting, a cyber security and data privacy consulting firm, where is the senior consultant on GDPR and CCPA. She is an IRCHSS PhD scholar at Dublin City University Business School, Ireland, where she is researching how organisational information privacy approaches. Previously, she spent fifteen years as Head of Information Security Risk Management at KBC Bank. Qualified as a Certified Information Systems Security Professional (CISSP) since 2001, she also holds a BSc from Trinity College in Information Systems, and an MBS from the Irish Management Institute in Business and Leadership.

Brid Murphy is an Assistant Professor in Accounting at Dublin City University Business School, Ireland. She is a fellow of the chartered accounting profession, and previously worked as an accounting practitioner in Ireland and overseas. Her research focuses on the impacts of technological and regulatory developments on the accounting profession and on accounting education innovations.

Marta Rocchi is Assistant Professor in Corporate Governance and Business Ethics at Dublin City University Business School and member of the Irish Institute of Digital Business. She holds a PhD from the University of Navarra, with a specialization on the ethics of finance. She was awarded the Society for Business Ethics Founders™ Award in 2016 as Emerging Scholar of the Society for Business Ethics, and the 1st prize *ex aequo* of the Ethics & Trust in Finance Global Prize of the Observatoire de la Finance in 2019. She teaches business and finance ethics, and her research focuses on virtue ethics in business and finance, the new perspectives of business ethics in the future of work, and the ethical dilemmas of the digital world. She published in prestigious journals in the business ethics field: Business Ethics Quarterly, Journal of Business Ethics, and Business Ethics: A European Review.

Lisa van der Werff is an Associate Professor of Organisational Psychology at DCU Business School and is Research Director of the Irish Institute of Digital Business. Her research focuses on trust development, workplace transitions and trust in technology. Dr van der Werff's inter-disciplinary work has been published in journals including the Academy of Management Journal, Journal of Management, Journal of Computer Information Systems, IEEE Transactions on Services Computing, and the Harvard Business Review. She sits on the editorial board of the Journal of Trust Research and is incoming president of FINT, an international network of trust scholars.

LIST OF FIGURES

LIST OF TABLES

Understanding Trust and Cloud Computing: An Integrated Framework for Assurance and Accountability in the Cloud

Theo Lynn, Lisa van der Werff, and Grace Fox

Abstract Trust is regularly cited as one the main barriers for increased adoption of cloud computing, however conceptualisations of trust in cloud computing literature can be simplistic. This chapter briefly introduces the trust literature including definitions and antecedents of trust. Following an overview of cloud computing, we discuss some of the cited barriers to trust in cloud computing, and proposed mechanisms for building trust in the cloud. We present a high-level framework for exploring assurance (trust building) and accountability (trust repair) in the cloud and call for a more integrated multi-stakeholder approach to trust research in this multi-faceted context.

T. Lynn (✉) • L. van der Werff • G. Fox
Irish Institute of Digital Business, DCU Business School,
Dublin, Ireland
e-mail: theo.lynn@dcu.ie; lisa.vanderwerff@dcu.ie; Grace.Fox@dcu.ie

© The Author(s) 2021   1
T. Lynn et al. (eds.), *Data Privacy and Trust in Cloud Computing*,
Palgrave Studies in Digital Business & Enabling Technologies,
https://doi.org/10.1007/978-3-030-54660-1_1

Keywords Trust • Cloud computing • Trust building • Trust repair • Assurance • Accountability

1.1   INTRODUCTION

Trust. A word that, while commonly used, is a complex concept that means different things to different people in different contexts. Technology is no different. "We don't trust the cloud" is a common phrase used to describe consumer or industry reluctance to adopt cloud computing. You will find it, or wording to the same effect, in numerous scholarly studies, industry surveys, and media, new and old. No matter what part of the economy, society, or world that you are in, you can find a report or survey suggesting that significant proportions of the public, businesses of all sizes, and the public sector do not or should not trust the cloud. Similarly, there are a myriad of, often conflicting, proposals and 'solutions' for overcoming trust issues in cloud computing. These include greater regulation, increased certification, stronger security, anonymity, trust by design, privacy by design, and so on. Indeed the importance of establishing trust in the cloud has been highlighted time and time again both in industry and academic discourse, with trust heralded as a solution to ease any concerns related to privacy and security on the cloud.

The objective of this book is to make some progress in teasing out what trust means in the context of cloud computing through a variety of lenses—psychology, law, ethics, information systems, and computing. The remainder of this chapter briefly introduces the trust literature including definitions and antecedents of trust. Next, we provide an overview of cloud computing and some of the reported trust-related barriers to cloud adoption and proposed solutions. Finally, we present a high-level framework for exploring assurance and accountability in the cloud.

1.2   TRUST

Trust is generally defined as a willingness to accept vulnerability based on positive expectations of another party (Rousseau et al. 1998). This definition has two critical elements—first, the psychological state of willingness to be vulnerable which represents a volitional choice or decision (van der Werff et al. 2019a). Second, there are positive expectations of another party, which refers to the influence of proximal antecedents or drivers of trust. Thus far, the trust literature has focused predominantly on a relatively small subset of proximal trust antecedents known as trustworthiness (Baer and Colquitt 2018). Trustworthiness is an aggregate perception of the characteristics of another party along three sub-dimensions: ability, integrity, and benevolence (Mayer et al. 1995). These concepts have been applied within the context of technology and appear regularly in the information systems literature (see van der Werff et al. 2018 for a review). This section will provide an overview of several potential antecedents of trust in cloud computing organised into two broad categories: knowledge based antecedents, including trustworthiness, and heuristic antecedents.

1.2.1 Knowledge Based Antecedents

The two aspects of trustworthiness most commonly studied in the trust in technology literature are ability and integrity. Ability or competence refers to a perception that the other party possesses the skills and knowledge to complete the tasks expected. This aspect of trustworthiness is readily applicable to perceptions of technology in terms of its performance levels including accuracy, capability and functionality (McKnight et al. 2011; Söllner et al. 2016). That is, *can* this cloud service do what I need it to do well? Integrity generally refers to the perception that another party adheres to a set of principles that the trustor finds acceptable, acts honestly and fulfils their promises (Mayer et al. 1995; McKnight et al. 1998). In the technology environment, this concept has typically been translated as a perception of reliability and consistency in performance. For instance, *will* this cloud service do what I need it to do *every* time I use it? In this setting in particular, the conceptualisation of integrity is expanded to integrate aspects of predictability and the extent to which it is possible to anticipate the other party's behaviour accurately (van der Werff et al. 2018). Interestingly, as they are applied in the computer science literature (see Chap. 7), these aspects of trustworthy cloud computing are sometimes portrayed as an objective feature of the technology rather than a more subjective user's perception of the technology as the original trust theory intended. This difference has particularly important implications in situations where the decision maker is not a technology expert and so subjective perceptions of trustworthiness are likely to differ significantly from any objective reality.

The third aspect of trustworthiness, benevolence, has received less attention in the cloud computing literature. As a perception of the extent to which another party will act in your best interests, benevolence incorporates aspects of agency and motivation into calculations of trustworthiness. Does the other party *want* to act in my best interests? At the moment, cloud services are not likely to act with either agency or motivation and benevolence perceptions have been applied in this context as a perception of alignment between user needs and the technology's purpose, helpfulness and responsiveness (McKnight et al. 2011; Söllner et al. 2016). However, while we may have some way to go before cloud services are automated to the point of agency, for many users anthropomorphisation of technology is common and perceptions of its motives and intentions are likely to play a role in trust decisions (Shank and DeSanti 2018).

1.2.2   Heuristic Antecedents

The use of knowledge based cues for trust is sometimes described as trust based on "good reasons" or rational decision making (Lewis and Weigert 1985, p. 970). However, a growing body of theoretical work and empirical evidence suggests that trust processes can be influenced by less rational antecedents and by beliefs about other related entities. The idea that such factors impact trust has gained traction over the last decade particularly in relation to trust in new or unknown other parties (e.g. Baer et al. 2018; Kramer and Lewicki 2010; McKnight et al. 1998) and trust in technology (e.g. McKnight et al. 2011). This section will briefly discuss four antecedents that may have a heuristic influence on trust in cloud computing: situational normality, aesthetics, structural assurances, and relational context.

The concept of situational normality was originally introduced to the trust literature by McKnight et al. (1998) who proposed that feeling like a situation was normal, familiar or as expected could be a powerful heuristic in building trust in unknown other parties. Since then, empirical evidence has repeatedly demonstrated the utility of situational normality as an antecedent of trust in organisations (Baer et al. 2018), e-commerce (Gefen 2000), recommendation agents (Komiak and Benbasat 2006) and software using speech (Lee 2010). The concept of situational normality is also readily observable in the context of cloud computing where cloud storage solutions integrate with other software on a user's personal computer to make the transition from personal to cloud storage as normal and un-noteworthy as possible.

A second heuristic influence on trust is aesthetics. This cue for trust relies on the halo effect which began as a concept in the social psychology literature to describe how immediately observable positive attributes such as physical attractiveness influence perceptions of other attributes. It has since been applied to the trust literature and used to explain everything from the outcomes of elections (Todorov et al. 2005) and new employees trust in organisations (Baer et al. 2018) to trust in websites (Cyr et al. 2010) and mobile commerce (Li and Yeh 2010). Regardless of the referent, the general principle of aesthetics cues is that other parties who are seen as aesthetically appealing are also likely to be seen as trustworthy, particularly in the early stages of a relationship.

Structural assurance is a cue for trust that is based less on a perception of the trust referent itself but more on a perception of the environment within which an interaction takes place. Kramer and Lewicki (2010) refer to this type of trust as rule based trust influenced by a perception that some form of checking or restraint in the environment will prevent another party from acting in a way that is not trustworthy. Again this concept, has proved useful in understanding trust in technology and evidence suggests that the effectiveness of regulatory and assurance systems can influence consumer trust in technology (e.g. Gefen and Pavlou 2006).

The final cue that has received attention in the literature also relates to the wider context of the trust relationship. Recent theory suggests that the immediate relational context plays a significant role in creating trust motivation or a desire to trust another party on the basis of the social function of the relationship (van der Werff et al. 2019a). In essence, if a technological artefact fulfils an important role for us in terms of depending on it to do something necessary, enjoying interacting with it or seeing it as being in line with our identity and personal values, we are more likely to trust it. Many relationships take place in a wider context or chain of interrelated parties. A growing body of evidence suggests that information about parties at another level in that chain can be used as a cue for trust (De Cremer et al. 2018; Lipponen et al. 2020) and that trust in one party can be transferred to referents at another level (Stewart 2003). It is likely in the technology context that information regarding other parties in a chain and the trust this information engenders can lead to trust in other parties.

1.3   CLOUD COMPUTING

Despite its ubiquity, cloud computing, as we know it today, is a recent phenomenon. It is hard to relate to the idea that when a company known for selling books online, Amazon, launched Amazon Web Services in 2006, it would help create a public cloud computing market worth nearly US$200 billion by 2019 (IDC 2019). In its most widely referenced definition, NIST define cloud computing as:

> ...model for enabling ubiquitous, convenient, on-demand network access to a shared pool of configurable computing resources (e.g., networks, servers, storage, applications, and services) that can be rapidly provisioned and released with minimal management effort or service provider interaction. This cloud model is composed of five essential characteristics, three service models, and four deployment models. (Mell and Grance, p. 2)

For the most part, the cloud model defined by Mell and Grance and the subsequent cloud reference architecture introduced by Liu et al. (2011) continue to be the basis of cloud computing industry. However, it would be wrong to say that cloud computing has not evolved. In particular, the emergence of the Internet of Things and Big Data, has led to the introduction and increasing adoption of a new service model, Function-as-a-Service, and two new computing paradigms, fog computing and edge computing (Lynn et al. 2017; Iorga et al. 2018). While further discussion is beyond the scope of this chapter, it is useful to be aware of these concepts and technology paradigms when considering trust and privacy issues, not only in this chapter but throughout the book. It is also important to note that these are not the only developments in cloud computing but the most influential at the time of writing. Table 1.1 below provides a brief definition of these some of the key concepts in cloud computing.

The essential characteristics of cloud computing, provide a wide range of benefits to businesses including increased infrastructure reliability and scalability (up and down), improved cashflow through reduced capital expenditure (CapEx) and operational expenditure (OpEx), as well providing competitive capabilities through increased agility, faster time-to-market, and new revenue streams (Lynn 2018). The induced effect for consumers is better quality of service and quality of experience, at lower or no financial cost. In the last two decades, advances in the coverage, speed, and reliability of global telecommunications networks has made the large

Table 1.1 Definitions of key concepts in cloud computing

Concept	Cloud essential characteristics	Source
On-demand self-service	A consumer can unilaterally provision computing capabilities, such as server time and network storage, as needed, automatically without requiring human interaction with each cloud service provider.	Mell and Grance (2011)
Broad network access	Capabilities are available over the network and accessed through standard mechanisms that promote use by heterogeneous thin or thick client platforms.	
Resource pooling	The cloud service provider's computing resources (e.g. storage, processing power, network bandwidth) are pooled to serve multiple consumers using a multi-tenant model, with different physical and virtual resources dynamically assigned and reassigned according to consumer demand.	
Rapid elasticity	Capabilities can be elastically provisioned and released, to scale rapidly outward and inward commensurate with demand.	
Measured service	Cloud systems automatically control and optimize resource use by leveraging a metering capability at some level of abstraction appropriate to the type of service.	
Cloud service models		
Software as a Service (SaaS)	The capability provided to the consumer is to use the provider's applications running on a cloud infrastructure and accessible by a client interface.	Mell and Grance (2011)
Platform as a Service (PaaS)	The capability provided to a consumer to deploy onto the cloud infrastructure consumer-created or acquired applications created using development technologies provided by the provider.	Mell and Grance (2011)
Infrastructure as a Service (IaaS)	The capability provided to the consumer to provision processing, storage, networks, and other fundamental computing resources to deploy and run arbitrary software.	Mell and Grance (2011)
Function as a Service (FaaS)	The capability provided to the consumer to execute lightweight, single purpose stateless functions that can be executed on demand, typically through an API, without consuming any resources until the point of execution.	Glikson et al. (2017) and Lynn (2018)

(*continued*)

Table 1.1 (continued)

Concept	Cloud essential characteristics	Source
Cloud deployment models		
Private cloud	Cloud infrastructure is provisioned for exclusive use by a single organization comprising multiple consumers. It may be owned, managed, and operated by the organization, a third party, or some combination of them, and it may exist on or off premises.	Mell and Grance (2011)
Community cloud	Cloud infrastructure is provisioned for exclusive use by a specific community of consumers from organizations that have shared concerns. It may be owned, managed, and operated by one or more of the organizations in the community, a third party, or some combination of them, and it may exist on or off premises.	Mell and Grance (2011)
Public cloud	Cloud infrastructure is provisioned for open use by the general public. It may be owned, managed, and operated by a business, academic, or government organization, or some combination of them. It exists on the premises of the cloud provider or their designated datacentre provider.	Mell and Grance (2011)
Hybrid cloud	Cloud infrastructure is a composition of two or more distinct cloud infrastructures (private, community, or public) that remain unique entities, but are bound together by standardized or proprietary technology that enables data and application portability.	Mell and Grance (2011)
Related computing paradigms		
Fog computing	Fog computing is a layered model for enabling ubiquitous access to a shared continuum of scalable computing resources. The model facilitates the deployment of distributed, latency-aware applications and services, and consists of fog nodes (physical or virtual), residing between smart end-devices and centralized (cloud) services.	Iorga et al. (2018)

(*continued*)

Table 1.1 (continued)

Concept	Cloud essential characteristics	Source
Edge computing	Edge computing is the network layer encompassing the end devices and their users, to provide, for example, local computing capability on a sensor, metering or some other devices that are network-accessible.	Iorga et al. (2018)
Dew computing	Dew computing is an on-premises computer software-hardware organization paradigm in the cloud computing environment where the on-premises computer provides functionality that is independent of cloud services and is also collaborative with cloud services.	Wang (2016)
Mist computing	Mist computing is an optional lightweight and rudimentary form of computing power that resides directly within the network fabric at the edge of that fabric, the fog layer closest to the smart end-devices, using microcomputers and microcontrollers to feed into fog computing nodes and potentially onward towards the cloud computing services.	Iorga et al. (2018)

scale outsourcing of information systems a reality. Consequently, more and more organisations are migrating from on-premise infrastructure to the cloud to focus on their core capabilities and to exploit potential IT efficiencies and business agility offered by the cloud (Kim 2009).

1.4   Trust Barriers to Cloud Adoption

Cloud computing is a form of outsourcing where organisations, and indeed albeit at a smaller scale, consumers, outsource some or all of their IT infrastructure (hardware, software, networks etc.) to one or more cloud service providers (CSP) on a metered basis. In return for fees, the CSP agrees to provide access to the cloud service at agreed service levels, typically contained in a Service Level Agreement (SLA).

Like all outsourcing, the decision to adopt cloud computing involves organisations assuming four main risks—relational, performance, compliance and regulatory, technological risks. Relational risk typically involves poor cooperation and opportunistic behaviour (Das and Teng 1996). As a by-product of both the on-demand nature of cloud computing and dominance of a relatively small number of hyperscale CSPs, standard form contracts are commonplace. Only the largest customers or those customers a CSP considers strategic, for example governments, have room to negotiate terms, or to develop a personal relationship with these providers. In the absence of a personal relationship, cloud computing relies largely on rule- or calculus-based trust, represented by these agreements. As will be discussed later in Chap. 2, not only do cloud computing contracts typically favour the service provider but cloud customers can find themselves locked-in from a technical perspective and dependent on the CSP for business continuity with important implications for trust.

Historically, performance risk has been the primary concern with cloud computing as evidenced by the focus of industry and scholars on service levels and SLAs. Clearly, availability and access are critical if one outsources IT infrastructure to the cloud. This is often further complicated by uncertainty related to the functioning of the cloud services, transparency on how service levels are calculated and of the underlying cloud systems and associated system data, and exceptions included in cloud contracts. Again, given the disparity in dependence and impact in the vendor-customer relationship, the risk of failure is significantly higher on the part of the customer.

The third risk, compliance and regulatory risk is where a customer fails to adhere to regulatory standards due to the provider's errors (Anderson et al. 2014). Increasingly but not exclusively, the primary barriers to cloud adoption, by organisations and consumers alike, relate to data, and more specifically the location, integrity, portability, security and privacy of data (Lynn et al. 2014; Leimbach et al. 2014; Eurostat 2016). Cloud computing is a largely location-independent technology and is built on a chain of service provision which is largely opaque to the customer. Data may be stored, processed, and transported across borders, and/or come in to contact with a wide range of partners, without the knowledge of the customer. Furthermore, CSPs, no matter what size are not immune from security vulnerabilities. Each service model, deployment model, and architecture, and combination and configuration thereof has its own discrete set of security issues. For SaaS models alone, Subashini and Kavitha (2011) identify 14 security elements that need to be considered independently of the PaaS and IaaS infrastructure upon which these are situated. At and within each layer, different parties may be responsible and accountable for the security of different elements. This is particularly pertinent in the context of data protection laws, such as the General Data Protection Regulation (GDPR), where misuse or mismanagement of data can result in significant fines and penalties, independent of the loss of reputation, and potential loss of corporate value associated with data and other security breaches (Goel and Shawky 2009).

Against this backdrop and in the absence of a personal relationship or knowledge, prospective customers and users of the cloud are faced with a relatively stark choice: To stay or go. The former involves assuming the risk laid out, relying on the contracts provided, and the competence, benevolence, and integrity of the CSP, while mitigating risks by other means, if possible or desirable. The alternative is to forego the benefits of the cloud altogether.

1.5   Existing Approaches to Overcoming Trust Barriers to Cloud Adoption

In addition to contracts, a variety of trust-building mechanisms have been proposed by policymakers, industry, and scholars. These include regulation, standardization, certification, communication, and technological innovation. For over a decade, the European Commission has sought to mitigate the impact of the risks outlined above through the activities leading to and from the 2012 European Cloud Strategy (European Commission 2012) and subsequent initiatives including the new European digital strategy, Shaping Europe's Digital Future (European Commission 2020). In addition to the GDPR, consumer protection regulations are in place to protect them from behaviour and contracts prejudicial to their consumer rights (see Chap. 2). Similarly, there have been numerous efforts to support standards not only for cloud system interoperability and data portability, but also for SLAs (see for example C-SIG-SLA 2014), however these are not mandatory. More recently, there has been a renewed focus on certification as a means of assurance.

Assurance involves expert practitioners evaluating an CSP against agreed criteria to improve the degree of confidence of intended users. In effect, this involves a cloud service provider redesigning their security and management processes to meet the requirements of a certification scheme, and then being audited by an independent third party to assess compliance periodically (Tecnalia 2016). This approach provides an opportunity for rule-based trust to develop and, in situations where the providers of the certification are trusted, the potential for trust transfer to occur. In a report for the European Commission published in 2018, Tecnalia identified over 20 such schemes, the most popular being compliance with ISO 27001; others included CSA Star, PCI-DSS, ENISA-CCM and the SOC (ISAE-3402) (Tecnalia 2016). A major limitation of the certification approach is the timeliness and the depth of the audit. In-depth audits may only take place every three years with light-touch reviews annually. Similarly, given the complexity of cloud computing, the level of detail that a certification or an auditor can go to is limited.

Three common methods are used to communicate trust in CSPs— website design, feedback mechanisms, and third party endorsements (Lynn et al. 2016). There is a substantial body of literature on the direct and indirect impact of visual website appearance on trust including colour choice and design symmetry which represent powerful heuristic cues for trust. However, aesthetic preferences in website design tend to vary across demographic characteristics and thus may have limited practical utility for CSPs trying to communicate trust (Cyr et al. 2010; Tuch et al. 2010). Feedback mechanisms or reputation systems are an increasingly popular alternative mechanism for communicating trust. As cloud and API marketplaces have emerged, such as Salesforce AppExchange, Microsoft Azure Marketplace and RapidAPI, so too have market-driven feedback systems within these marketplaces. Ratings, reviews, and vendor ecosystem status all act as a signal to consumers that the vendor has an incentive to behave in an appropriate manner and that they have been informally certified by previous consumers (Pavlou and Gefen 2004). Again, these mechanisms are likely to impact trust by providing a level of structural assurance and cues regarding the rules governing trustworthy behaviour. Independently of the cloud sector, a plethora of general reputation and review systems, such as Feefo and Trust Pilot, have emerged in recent years that seek to provide prospective customers, both business-to-business (B2B) and business-to-consumer (B2C), with similar signals on an independent basis by aggregating ratings, surveys and reviews (Banerjee et al. 2020). Increasingly, these are integrated not only in to a vendor's website but into search engine ranking algorithms, providing additional incentives for vendors to behave. Notwithstanding their widespread and increasing use, feedback and reputation systems have been criticised for their vulnerability to false, manipulated or biased feedback (Sabater and Sierra 2005).

A third approach to communicating trust in CSPs involves the use of assurance seals or trustmarks that combine certification and communication to dispel consumer concerns about risk and communicate adherence with best practice, a code of conduct, or certification scheme using a third-party mark or symbol (Aiken and Boush 2006). Like certification, trustmark holders are typically subject to periodic third party verification. However, in addition to recognition and lack of information depth, trustmarks suffer from the same limitations as certification in general. They have been criticised for reliance on human intervention, limited scope, timeliness, lacking warranties, and subject to co-optation risk (Aiken et al. 2003).

Technological innovation to build trust in cloud computing largely revolves around designing clouds that meet the three pillars of trustworthy computing—security and privacy, reliability, and business integrity (Mundie et al. 2002). Chapter 7 discusses this topic in detail. It is important to note, however, that technical innovation in trustworthy computing overwhelmingly focuses on the first two pillars, security and privacy, and reliability. Research on the former focuses on the provision of effective attack resilient systems, typically using encryption techniques of increasing strength and complexity. Reliability research focuses on the design, monitoring, and measurement of highly reliable systems. Both domains are largely hidden from end-users. Business integrity is more nuanced and suffers from a lack of inter-disciplinary research. As such, it focuses largely on monitoring key service level metrics and ranking services based on this data. One of the main limitations of purely technological approaches, is that by and large, customers are human. Their decisions to trust are based on a vast array of conscious and subconscious signals that are often forgotten about in purely technological approaches and solutions.

In attempt to address this gap and marry the various approaches to mitigating trust issues in cloud computing, we have previously proposed an active dynamic online trust label (Lynn et al. 2014; Lynn et al. 2016; Emeakaroha et al. 2016; van der Werff et al. 2019b). Inspired by nutritional labels, these labels present consumers with corporate information, policies, and historic and near real-time service level metrics based on data from CSP monitoring systems (Emeakaroha et al. 2016). The system can allow for third party independent certification and could allow for corporate attestation using digital signatures. Based on an experimental study with 227 business decision makers, the proposed cloud trust label communicated trustworthiness effectively (van der Werff et al. 2019b). While these results are promising, such a system requires widespread support to be effective. Until then, it remains an academic exercise.

1.6   ASSURANCE AND ACCOUNTABILITY FRAMEWORK

In general, mechanisms to build trust in cloud computing fall in to two main categories—assurance and accountability. Standards, certification, and communication strategies seek to assure the consumers by providing cues of CSP competence, integrity, and benevolence, and to some extent consistency. Regulation and contractual mechanisms seek to hold CSPs accountable in the event of a trust violation. A key problem is that these initiatives are currently highly fragmented, with multiple initiatives by as many stakeholders, but no particular comprehensive, coordinated, and holistic framework of activity that provides direction for policy makers, users, cloud service providers, and indeed researchers.

Figure 1.1 below presents an integrated multi-stakeholder framework for assurance and accountability for cloud-based trust building. It extends the chain of accountability concept first proposed by Pearson and Wainwright (2013) to provide transparency and clarity on liability in the event of a data breach in the cloud. While Pearson and Wainwright (2013) envisaged a set of mechanisms for mitigating risk (preventative controls), monitoring and identifying risk and policy violations (detective controls), and providing redress (corrective controls), their approach is largely built on calculative trust-based model whereby accountability is both quantitative and absolute. The goal is to eliminate distrust or mitigate the negative impact of a trust violation. In effect, it is an *ab initio* pre-emptive trust repair approach.

In contrast, we propose, a more positive approach couched in theories of trust building and repair. The focus is on trust building mechanisms; trust repair mechanisms only initiate when a trust violation occurs. Based on our work in Lynn et al. (2014), we suggest that cloud consumers should have control of their data, how it is used, where it is used, and who should use it, and this should be auditable by all involved. They should have a say, if they want it, but as a default standard declarations should be weighed towards the best interests of the consumer, and neither prejudicial to consumer rights, nor contrary to government policy. As such, we propose that in addition to preventative controls, there are declarative controls where all parties can declare their policies and expectations irrespective of contracts or policies which seek to circumvent local laws and regulation. Furthermore, there are confirmative controls that report and alert stakeholders that these policies and expectations are being met. In this way, trust is not only being built on the basis on rules and transactions, but proactive mechanisms are in place so that knowledge-based

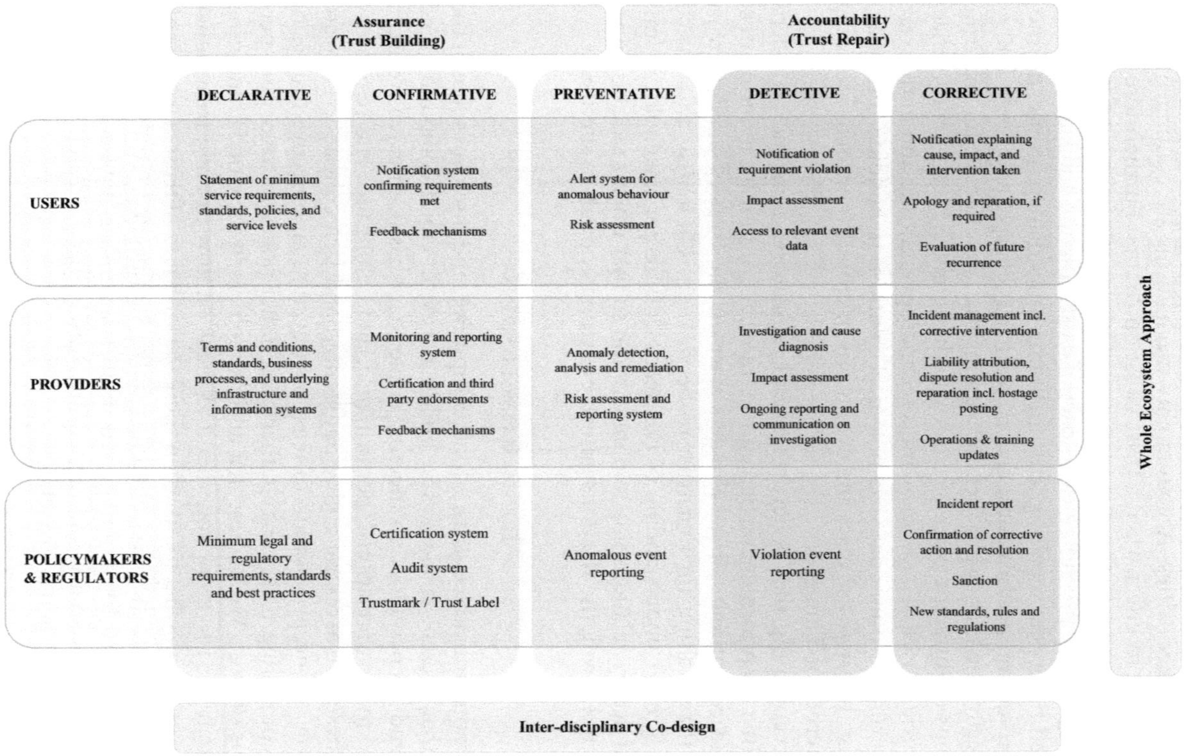

	Assurance (Trust Building)			**Accountability (Trust Repair)**	
	DECLARATIVE	CONFIRMATIVE	PREVENTATIVE	DETECTIVE	CORRECTIVE
USERS	Statement of minimum service requirements, standards, policies, and service levels	Notification system confirming requirements met; Feedback mechanisms	Alert system for anomalous behaviour; Risk assessment	Notification of requirement violation; Impact assessment; Access to relevant event data	Notification explaining cause, impact, and intervention taken; Apology and reparation, if required; Evaluation of future recurrence
PROVIDERS	Terms and conditions, standards, business processes, and underlying infrastructure and information systems	Monitoring and reporting system; Certification and third party endorsements; Feedback mechanisms	Anomaly detection, analysis and remediation; Risk assessment and reporting system	Investigation and cause diagnosis; Impact assessment; Ongoing reporting and communication on investigation	Incident management incl. corrective intervention; Liability attribution, dispute resolution and reparation incl. hostage posting; Operations & training updates
POLICYMAKERS & REGULATORS	Minimum legal and regulatory requirements, standards and best practices	Certification system; Audit system; Trustmark / Trust Label	Anomalous event reporting	Violation event reporting	Incident report; Confirmation of corrective action and resolution; Sanction; New standards, rules and regulations

Whole Ecosystem Approach

Inter-disciplinary Co-design

Fig. 1.1 An integrated multi-stakeholder framework for building and repairing trust in cloud computing based on assurance and accountability

trust is being built between all stakeholders. These two assurance based controls are necessities. Accountability mechanisms are contingent; they only come in to effect when a trust violation occurs. Furthermore, when initiated, these mechanisms are not mere objective features of the system but recognise the psychological impact of trust violation and largely follow accepted theory for repairing trust including immediate response, diagnosis, intervention performance, and evaluation (Gillespie and Dietz 2009). Specifically, the framework includes actions that are effective for repairing violations of different types of trust, whether competence-, benevolence- or integrity-based. The framework is technology-agnostic and in this way, can not only accommodate technological solutions to building and repairing trust, but new use cases and evolutions of cloud computing including the Internet of Things.

By recognising that policymakers and regulators, users and providers, have different priorities and perceptions of what trust means in the context of cloud computing, all stakeholders start on the basis of building trust rather than waiting for that trust to be violated. Ultimately, this should lead to greater understanding of the needs of different stakeholders, longer and deeper relationships, and innovation so that when a violation does occur, and it will, the relationship will be strong enough to survive.

1.7   CONCLUSIONS

This chapter introduces trust, cloud computing, and discusses some of the issues that present challenges to building trust in cloud computing, and wider and deeper adoption thereof. While there has been extensive work done to mitigate relational, performance, and compliance and regulatory risks, these initiatives are highly fragmented and lack cohesion. They are based on a conceptualisation of trust portrayed as an objective feature of cloud computing technology rather than either policymaker or user perceptions of trust. We suggest that all stakeholders in the cloud computing ecosystem need to come together and focus on how to build trust rather than focusing on what to do when there is a violation of trust, a reposition to assurance first, then accountability only when needed. To this end, we reiterate the need for an integrated multi-stakeholder approach to assurance and accountability, and related inter-disciplinary research to support the adoption of such approaches.

REFERENCES

Aiken, D., Osland, G., Liu, B., & Mackoy, R. (2003). Developing Internet Consumer Trust: Exploring Trustmarks as Third-Party Signals. *Marketing Theory and Applications, 14*, 145–146.

Aiken, K. D., & Boush, D. M. (2006). Trustmarks, Objective-Source Ratings, and Implied Investments in Advertising: Investigating Online Trust and the Context-Specific Nature of Internet Signals. *Journal of the Academy of Marketing Science, 34*(3), 308–323.

Anderson, S. W., Christ, M. H., Dekker, H. C., & Sedatole, K. L. (2014). The Use of Management Controls to Mitigate Risk in Strategic Alliances: Field and Survey Evidence. *Journal of Management Accounting Research, 26*(1), 1–32.

Baer, M., & Colquitt, J. A. (2018). Moving Toward a More Comprehensive Consideration of the Antecedents of Trust. In R. H. Searle, A. M. Neinaber, & S. B. Sitkin (Eds.), *Routledge Companion to Trust* (pp. 163–182). Abingdon: Routledge.

Baer, M. D., Van Der Werff, L., Colquitt, J. A., Rodell, J. B., Zipay, K. P., & Buckley, F. (2018). Trusting the "Look and Feel": Situational Normality, Situational Aesthetics, and the Perceived Trustworthiness of Organizations. *Academy of Management Journal, 61*(5), 1718–1740.

Banerjee, A., Ries, J. M., & Wiertz, C. (2020). The Impact of Social Media Signals on Supplier Selection: Insights from Two Experiments. *International Journal of Operations & Production Management.* https://doi.org/10.1108/IJOPM-05-2019-0413.

C-SIG-SLA. (2014). Cloud Service Level Agreement Standardisation Guidelines. Retrieved from http://ec.europa.eu/newsroom/dae/document.cfm?action=display&doc_id=6138

Cyr, D., Head, M., & Larios, H. (2010). Colour Appeal in Website Design Within and Across Cultures: A Multi-method Evaluation. *International Journal of Human-Computer Studies, 68*(1), 1–21.

Das, T. K., & Teng, B. S. (1996). Risk Types and Inter-firm Alliance Structures. *Journal of Management Studies, 33*(6), 827–843.

De Cremer, D., Van Dijke, M., Schminke, M., De Schutter, L., & Stouten, J. (2018). The Trickle-Down Effects of Perceived Trustworthiness on Subordinate Performance. *Journal of Applied Psychology, 103*(12), 1335.

Emeakaroha, V. C., Fatema, K., van der Werff, L., Healy, P., Lynn, T., & Morrison, J. P. (2016). A Trust Label System for Communicating Trust in Cloud Services. *IEEE Transactions on Services Computing, 10*(5), 689–700.

European Commission. (2012). *Communication from the Commission to the European Parliament, the Council, the European Economic and Social Committee and the Committee of the Regions.* Unleashing the Potential of Cloud Computing in Europe. COM(2012) 529 Final.

European Commission. (2020). Shaping Europe's Digital Future. Retrieved from https://ec.europa.eu/info/sites/info/files/communication-shaping-europes-digital-future-feb2020_en_4.pdf

Eurostat. (2016). Archive: Cloud Computing—Statistics on the Use by Enterprises—2016 Data. Retrieved from https://ec.europa.eu/eurostat/statistics-explained/index.php?title=Archive:Cloud_computing_-_statistics_on_the_use_by_enterprises_-_2016_data

Gefen, D. (2000). E-commerce: The Role of Familiarity and Trust. *Omega, 28*(6), 725–737.

Gefen, D., & Pavlou, P. (2006). The Moderating Role of Perceived Regulatory Effectiveness of Online Marketplaces on the Role of Trust and Risk on Transaction Intentions. *ICIS 2006 Proceedings*, 81.

Gillespie, N., & Dietz, G. (2009). Trust Repair After an Organization-Level Failure. *Academy of Management Review, 34*(1), 127–145.

Glikson, A., Nastic, S., & Dustdar, S. (2017, May). *Deviceless Edge Computing: Extending Serverless Computing to the Edge of the Network.* Proceedings of the 10th ACM International Systems and Storage Conference, pp. 1-1.

Goel, S., & Shawky, H. A. (2009). Estimating the Market Impact of Security Breach Announcements on Firm Values. *Information & Management, 46*(7), 404–410.

IDC. (2019). *Worldwide Public Cloud Services Spending Guide.* Framingham, MA: IDC.

Iorga, M., Feldman, L., Barton, R., Martin, M. J., Goren, N. S., & Mahmoudi, C. (2018). *Fog Computing Conceptual Model.* (No. Special Publication (NIST SP)-500-325).

Kim, W. (2009). Cloud Computing: Today and Tomorrow. *Journal of Object Technology, 8*(1), 65–72.

Komiak, S. Y., & Benbasat, I. (2006). The Effects of Personalization and Familiarity on Trust and Adoption of Recommendation Agents. *MIS Quarterly*, 941–960.

Kramer, R. M., & Lewicki, R. J. (2010). Repairing and Enhancing Trust: Approaches to Reducing Organizational Trust Deficits. *Academy of Management Annals, 4*(1), 245–277.

Lee, E. J. (2010). The More Humanlike, the Better? How Speech Type and Users' Cognitive Style Affect Social Responses to Computers. *Computers in Human Behavior, 26*(4), 665–672.

Leimbach, T., Hallinan, D., Bachlechner, D., Weber, A., Jaglo, M., Hennen, L., Nielsen, R. O., Nentwich, M., Strauss, S., Lynn, T., & Hunt, G. (2014). *Potential and Impacts of Cloud Computing Services and Social Network Websites.* Publication of Science and Technology Options Assessment.

Lewis, J. D., & Weigert, A. (1985). Trust as a Social Reality. *Social Forces, 63*(4), 967–985.

Li, Y. M., & Yeh, Y. S. (2010). Increasing Trust in Mobile Commerce Through Design Aesthetics. *Computers in Human Behavior, 26*(4), 673–684.

Lipponen, J., Kaltiainen, J., van der Werff, L., & Steffens, N. K. (2020). Merger-Specific Trust Cues in the Development of Trust in New Supervisors During an

Organizational Merger: A Naturally Occurring Quasi-Experiment. *The Leadership Quarterly, 31*(4), 101365.

Liu, F., Tong, J., Mao, J., Bohn, R., Messina, J., Badger, L., & Leaf, D. (2011). NIST Cloud Computing Reference Architecture. *NIST Special Publication, 500,* 292.

Lynn, T. (2018). Addressing the Complexity of HPC in the Cloud: Emergence, Self-Organisation, Self-Management, and the Separation of Concerns. In *Heterogeneity, High Performance Computing, Self-Organization and the Cloud* (pp. 1–30). Cham: Palgrave Macmillan.

Lynn, T., Healy, P., McClatchey, R., Morrison, J., Pahl, C., & Lee, B. (2014). The Case for Cloud Service Trustmarks and Assurance-as-a-Service. preprint arXiv:1402.5770.

Lynn, T., Rosati, P., Lejeune, A., & Emeakaroha, V. (2017, December). *A Preliminary Review of Enterprise Serverless Cloud Computing (Function-as-a-Service) Platforms.* 2017 IEEE International Conference on Cloud Computing Technology and Science (CloudCom) (pp. 162–169). IEEE.

Lynn, T., Van Der Werff, L., Hunt, G., & Healy, P. (2016). Development of a Cloud Trust Label: A Delphi Approach. *Journal of Computer Information Systems, 56*(3), 185–193.

Mayer, R. C., Davis, J. H., & Schoorman, F. D. (1995). An Integrative Model of Organizational Trust. *Academy of Management Review, 20*(3), 709–734.

McKnight, D. H., Carter, M., Thatcher, J. B., & Clay, P. F. (2011). Trust in a Specific Technology: An Investigation of Its Components and Measures. *ACM Transactions on Management Information Systems (TMIS), 2*(2), 12.

McKnight, D. H., Cummings, L. L., & Chervany, N. L. (1998). Initial Trust Formation in New Organizational Relationships. *Academy of Management Review, 23*(3), 473–490.

Mell, P., & Grance, T. (2011). The NIST Definition of Cloud Computing (Draft). *NIST Special Publication, 800,* 145.

Mundie, C., de Vries, P., Haynes, P., & Corwine, M. (2002). Trustworthy Computing-Microsoft White Paper. Microsoft Corporation, October.

Pavlou, P. A., & Gefen, D. (2004). Building Effective Online Marketplaces with Institution-based Trust. *Information Systems Research, 15*(1), 37–59.

Pearson, S., & Wainwright, N. (2013). An Interdisciplinary Approach to Accountability for Future Internet Service Provision. *International Journal of Trust Management in Computing and Communications, 1*(1), 52–72.

Rousseau, D. M., Sitkin, S. B., Burt, R. S., & Camerer, C. (1998). Not So Different After All: A Cross-Discipline View of Trust. *Academy of Management Review, 23*(3), 393–404.

Sabater, J., & Sierra, C. (2005). Review on Computational Trust and Reputation Models. *Artificial Intelligence Review, 24*(1), 33–60.

Shank, D. B., & DeSanti, A. (2018). Attributions of Morality and Mind to Artificial Intelligence after Real-World Moral Violations. *Computers in Human Behavior, 86,* 401–411.

Söllner, M., Hoffmann, A., & Leimeister, J. M. (2016). Why Different Trust Relationships Matter for Information Systems Users. *European Journal of Information Systems, 25*(3), 274–287.

Stewart, K. J. (2003). Trust Transfer on the World Wide Web. *Organization Science, 14*(1), 5–17.

Subashini, S., & Kavitha, V. (2011). A Survey on Security Issues in Service Delivery Models of Cloud Computing. *Journal of Network and Computer Applications, 34*(1), 1–11.

Tecnalia. (2016). Certification Schemes for Cloud Computing. Retrieved from https://op.europa.eu/en/publication-detail/-/publication/3df22a89-1238-1 1e9-81b4-01aa75ed71a1/language-en

Todorov, A., Mandisodza, A. N., Goren, A., & Hall, C. C. (2005). Inferences of Competence from Faces Predict Election Outcomes. *Science, 308*(5728), 1623–1626.

Tuch, A. N., Bargas-Avila, J. A., & Opwis, K. (2010). Symmetry and Aesthetics in Website Design: It's a Man's Business. *Computers in Human Behavior, 26*(6), 1831–1837.

van der Werff, L., Legood, A., Buckley, F., Weibel, A., & de Cremer, D. (2019a). Trust Motivation: The Self-Regulatory Processes Underlying Trust Decisions. *Organizational Psychology Review, 9*(2-3), 99–123.

van der Werff, L., Fox, G., Masevic, I., Emeakaroha, V. C., Morrison, J. P., & Lynn, T. (2019b). Building Consumer Trust in the Cloud: An Experimental Analysis of the Cloud Trust Label Approach. *Journal of Cloud Computing, 8*(1), 6.

van der Werff, L., Real, C., & Lynn, T. (2018). Individual Trust and the Internet. In R. H. Searle, A. M. I. Nienaber, & S. B. Sitkin (Eds.), *The Routledge Companion to Trust* (pp. 391–407). Abingdon: Routledge.

Wang, Y. (2016). Definition and Categorization of Dew Computing. *Open Journal of Cloud Computing (OJCC), 3*(1), 1–7.

CHAPTER 2

Dear Cloud, I Think We Have Trust Issues: Cloud Computing Contracts and Trust

Theo Lynn

Abstract Cloud computing is the dominant paradigm in modern computing, used by billions of Internet users worldwide. It is a market dominated by a small number of hyperscale cloud service providers. The overwhelming majority of cloud customers agree to standard form click-wrap contracts, with no opportunity to negotiate specific terms and conditions. Few cloud customers read the contracts that they agree to. It is clear that contracts in cloud computing are primarily an instrument of control benefiting one side, the cloud service provider. This chapter provides an introduction to the relationship between psychological trust, contracts and contract law. It also offers an overview of the key contract law issues that arise in cloud computing and introduces some emerging paradigms in cloud computing and contracts.

Keywords Contract law • Terms of service • Cloud computing • Cloud contracts • Trust

T. Lynn (✉)
Irish Institute of Digital Business, DCU Business School, Dublin, Ireland
e-mail: theo.lynn@dcu.ie

© The Author(s) 2021
T. Lynn et al. (eds.), *Data Privacy and Trust in Cloud Computing*,
Palgrave Studies in Digital Business & Enabling Technologies,
https://doi.org/10.1007/978-3-030-54660-1_2

2.1   INTRODUCTION

Since the 1990s, outsourcing information systems has been a staple of business strategists. They argue that firms should focus on their core competencies and outsource all other activities to optimise resource allocation (Lambert and Peppard 2013). The emergence of what the International Data Corporation (IDC 2013) term the 'Third IT Platform' comprising cloud computing, social media, mobile and big data/analytics technologies has accelerated the outsourcing of critical information systems.

Cloud computing has emerged as the dominant computing paradigm of the twenty-first century. It is defined as a "a model for enabling ubiquitous, convenient, on-demand network access to a shared pool of configurable computing resources that can be rapidly provisioned and released with minimal management effort or service provider interaction" (Mell and Grance 2011, p. 2). Enterprise cloud IT expenditure consistently exceeds non-cloud expenditure for enterprises of all sizes (IDC 2020). Despite the ubiquity of cloud computing, a small number of hyperscale cloud service providers (CSPs) dominate the public cloud market. In 2018, five companies accounted for nearly 77% of the global Infrastructure-as-a-Service (IaaS) market—Amazon, Alibaba, Google, IBM, and Microsoft (Gartner 2019). As cloud computing is the basis of the Internet, including most social networking sites, search engines, and mobile applications, over 3.6 billion Internet users rely on cloud computing for one or more services. For firms, the benefits of increased IT efficiencies, agility, and scalability (both up and down) must be weighed against the risks that these technologies pose to firm performance, relationships, and compliance (Lynn and Rosati 2017). Similarly, for individuals, they must weigh up the utility of the functionality they receive from the applications that they subscribe to, and the risk to their data privacy.

For the most part, cloud computing contracts are standard form and click-wrap in nature, with only the largest corporate and government customers in a position to negotiate tailored terms and conditions (Bradshaw et al. 2013). As part of this click-wrap procedure and before payment, prospective clients are typically presented with either (1) a scrollable agreement, or (2) a link to a webpage or downloadable agreement, based on the specific cloud service and configuration that they have selected. They are encouraged to review the text of the agreement, and asked to communicate assent to the terms and conditions by clicking on an interactive 'I agree' button. For the overwhelming majority of CSP clients, these click-wrap contracts are standardised wholly electronic contracts giving the clients little or no opportunity to negotiate specific terms and conditions. As such, the vast majority of cloud customers need to balance the tension of the advantages of the cloud against the disadvantages of boilerplate terms and conditions designed by global firms with legal resources several orders of magnitude greater than even the largest law firms, as well as the perceived and actual loss of control. In the absence of alternatives, cloud customers may feel that they have no choice but to rely on these contracts to eliminate distrust or mitigate the negative impact of a trust violation, in effect a form of what Lewicki et al. (2006) refer to as calculative trust. Similarly, they may simply agree to the terms and conditions as an anxiety avoidance mechanism (Weber et al. 2004; van der Werff et al. 2019). Either way, a trust issue arises.

This chapter provides an overview of common terms and conditions in general form cloud computing click-wrap contracts. To avoid repetition, we assume the general definitions of trust and cloud computing presented in Chap. 1. The remainder of the chapter is organised as follows. Following a brief discussion on the theoretical relationship between trust, contracts and contract law, the structure of cloud computing contracts is introduced. This is followed by an overview and discussion of the key terms and conditions in cloud computing contracts and the issues that these present. Then, we discuss briefly how the nature of cloud computing and contracts are evolving before concluding with a brief discussion of the trust implications resulting from these issues.

2.2   Trust, Contracts and Contract Law

Trust and distrust are inextricably linked to the moral and legal underpinnings of Anglo-American contract law. The purpose of this chapter is not to justify trust as a theoretical building block of contract law but rather outline contractual issues in cloud computing and how these might impact trust in cloud computing and CSPs. However, understanding the relationship between trust, contracts and contract law, even at a high level, may provide insights in to the role of cloud computing contracts play in the relationship between CSPs and their clients. At the core of a contract is a promise where "a person invites another to trust, and to break a promise is to abuse that trust" (Bellia Jr 2002, p. 25). But, what is the nature of this trust? And what is the relationship between trust, contracts, and contract law?

As discussed in Chap. 1, psychologists suggest that when we trust someone, we accept vulnerability based on positive expectations of the future behaviour of that party (Rousseau et al. 1998). Inherent in this trust, is the assumption that the other party (1) possesses the necessary skills and capabilities to deliver on the promise (ability); (2) has the trustor's interests at heart (benevolence); and (3) will adhere to a set of mutually acceptable principles for behaviour (integrity) (Mayer et al. 1995).

Kimel (2001) suggests that while promises draw on the same reliance and expectation of fulfilment that exists in personal trust, contracts are different than promises and exist outside of the framework of personal relationships. He argues that contracts, in fact, undermine the concept of psychological trust and human relationships and, in effect, exist as a substitute to trust (Kimel 2001). In contrast, Bellia Jr (2002) argues that the intrinsic value of a promise, regardless of enforceability, does not lie in its capacity to reinforce trust relationships but rather in the knowledge that certain promises need to be enforced. Similarly, Lumineau (2017) posits that a lack of trust does not necessarily signify distrust, and indeed argues that both trust and distrust can result in positive and negative outcomes.

Legal theorists use similar constructs to argue why the law should enforce a contract. Trust is conceptualised in a number of different ways in contract law theory. For example, autonomy theory argues that the enforcement threat in contract law exists to enhance the freedom of the promisor and respects the trust of the promise, while welfare-economic theorists argue it exists to maximise individual or social well-being (Bellia Jr 2002). In reality, contract law exists to perform a variety of trust-related functions including enabling parties to make and enforce a promise, avoid conflicts, and regulate coordination and cooperation between them (Bellia Jr 2002).

2.3   The Form of General Cloud Computing Contracts

The contractual relationship between CSPs, their clients, and crucially their clients' end users, are typically set out in a standard form click-wrap contract comprising the following four components:

- Terms of Service (TOS)—the TOS set out the provisions that define and regulate the overall relationship between a CSP and the client.

- Service Level Agreement (SLA)—the SLA details the level of service to be provided, often in the form of specific quality of service (QoS) metrics, and the mechanisms for auditing service delivery and QoS, and compensating clients for underperformance.
- Acceptable Use Policy (AUP)—sometimes called a 'fair use policy', the AUP is a policy to protect CSPs from the actions of clients, and in the case of enterprise clients, their end users, by detailing prohibited uses of the contracted cloud service.
- Privacy Policy—this details the CSP's policy for handling and protecting personal data, in line with data protection law requirements.

Click-wrap contracts are part of a common cloud service subscription procedure made over the Internet. As part of this procedure and before payment, prospective clients are typically presented with either (1) a scrollable agreement, or (2) a link to a webpage or downloadable agreement, based on the specific cloud service and configuration that they have selected. They are encouraged to review the text of the agreement, and asked to communicate assent to the terms and conditions by clicking on an interactive 'I agree' button. For the overwhelming majority of CSP clients, these click-wrap contracts are standardised wholly electronic contracts giving the clients little or no opportunity to negotiate specific terms and conditions. Bradshaw et al. (2013) notes three distinctions within cloud computing contracts—(1) free vs paid services, (2) US v EU jurisdictions, and (3) IaaS v Software-as-a-Service (SaaS). First, they note that some terms and conditions for paid services are more likely to be open to negotiation depending on the bargaining power of the prospective client e.g. large multinational corporations or Governments. In these cases, depending on the standing of the CSP in the market or the specific segment, there may be a relationship of interdependence rather than dependence (McKnight et al. 2002). This is particularly evident cloud application and API marketplaces (Paulsson et al. 2020). Second, contracts offered under US law have more extensive disclaimers of warranty and limitations of liability than those offered under European Union (EU) law (Bradshaw et al. 2013). Thirdly, terms and conditions offered by IaaS providers would seem to be more similar than those offered by SaaS providers (Bradshaw et al. 2013).

In contract law, the so-called 'informed minority' hypothesis has been used to justify the avoidance of regulation of standard form contracts (Bakos et al. 2014). This hypothesis posits that there is generally a significant number of informed consumers in any given market to make an informed decision on the terms of a standard-form contract, and that while a substantial number, if not the majority, of consumers may remain uninformed, the former is of significant size to discipline abuse by the market (Schwartz and Wilde 1978). Extant research has found that Internet users consistently do not read such click-wrap contracts. For example, Bakos et al. (2014) found only 0.2% shoppers access a product's end-user license agreement for at least one second. As mentioned earlier, this may be due to a combination of dependency and anxiety avoidance on behalf of the cloud consumer (Weber et al. 2004; van der Werff et al. 2019). To counter this, policymakers have sought to mandate both disclosure of terms and conditions, and acceptance (van der Wees et al. 2014). Notwithstanding this, research suggests accessibility and mandatory acceptance do not result in significant increases in reading click-wrap contracts, and even those who have read the contracts, do not change their decision or behaviour (Marotta-Wurgler 2012). While there is a paucity of similar research on firm behaviour with regards to click-wrap contracts, it is likely to be similar, particularly for smaller organisations. The reality is accepting click-wrap contracts has become a habitual and an inevitable part of cloud computing. By not reading the terms and conditions of these click-wrap agreements, there is no incentive for CSPs to provide anything more than the minimum legal requirements. As such, most cloud contracts are extremely one-sided (Bradshaw et al. 2013) and would not seem to be subject to the informed minority rule (Schwartz and Wilde 1978).

The enforceability of electronic click-wrap agreements has been upheld by courts worldwide for both business-to-consumer and business-to-business transactions, and for paid and free services, tending towards supporting the position of the service provider (see for example, *Rudder v Microsoft Corp, Caspi v Microsoft Network*, and *El Majdoub v CarsOnTheWeb. Deutschland GmbH*). The main arguments are both freedom of contract arguments i.e. that clients have the opportunity to make themselves familiar with the terms and conditions and that they provide consent, and economic arguments i.e. rendering click-wrap contracts ineffectual, even though one-sided, would disrupt e-commerce and not be in the public interest.

2.4   COMMON CHALLENGES AND ISSUES IN GENERAL CLOUD COMPUTING CONTRACTS

2.4.1   Choice of Law

By definition, cloud computing is a distributed model. Data can be, and most likely will be, stored and processed across multiple data centres, potentially in different jurisdictions, and even where stored and processed in one jurisdiction, may be transferred across borders and accessed in different jurisdictions. It is possible that the provider and the end user are unaware of where the data is processed. For enterprise clients, the TOS increasingly allow data residency in a specific region; a region may be a country or a larger area such as the European Union. While consumer cloud users may not have that choice, with the transposition of the General Data Protection Regulation (Directive 95/46/EC) (GDPR), CSPs typically store European data within the EU for compliance reasons. Notwithstanding this, a recent survey of 322 cloud TOS and privacy policies, suggested that 267 CSPs indicated that the US was the preferred jurisdiction, and specifically Californian law (Martic 2017).

Chapter 3 will discuss jurisdictional issues in greater detail, however it is important to highlight that choice of law can favour one side or the other in a cloud contract. For example, EU law does not allow the exclusion or limitation of liability to the same extent that US law might, and similarly the GDPR introduces significant responsibilities and penalties on data controllers and processors. Courts will consider a number of factors when deciding on the actual jurisdiction for a cloud contract including: (1) the choice of law in the TOS; (2) the nature and quality of the CSP's commercial activity in the jurisdiction; (3) whether the CSP is actively aware that they are making sales to client resident in a particular jurisdiction; (4) the jurisdiction that clients are resident or domiciled in; (5) the location where the cloud service is consumed; (6) the location whether the data is stored and processed; (7) the location of the CSP's offices; and (8) whether the CSP markets or solicits business in a given jurisdiction. If the answer to one or more of these questions is affirmative, a court may enforce jurisdiction.

In Europe, CSPs and enterprise clients often seek to use and rely on standard contract clauses, so-called EU model clauses, to manage data transfer outside the EU. However the applicability of these have been challenged in the recent case of *Data Protection Commissioner v Facebook Ireland (Schrems II)*. The judgment for this case was delivered in July

2020 with the CJEU largely following the advice of the Advocate General i.e. that model clauses should not be invalidated and that reliance on such clauses requires firms undertake additional measures to assure compliance. However, the CJEU, somewhat unexpectantly, decided to examine and rule the EU-U.S. Privacy Shield framework invalid thus requiring organisations relying on this mechanism to urgently consider and put in place alternatives.

2.4.2   Service Level Agreements and Limitation of Liability

The SLA outlines the CSP's commitments on availability, reliability, and performance levels for the specific cloud service contracted. These are typically presented as quantifiable targets for the standard of service, how such targets are calculated, mechanisms for auditing service delivery, and the level and procedure for compensation in the event of under-performance (Leimbach et al. 2014). The exclusions in SLAs can be quite broad and typically include an amount of scheduled downtime per annum (e.g. for maintenance) but also factors outside of the CSP's immediate control. Again, these are rarely negotiable on the grounds that the traditional cloud computing business model is based on multi-tenancy and commoditisation; negotiation is only available for those with significant bargaining power (Weber and Staiger 2014; Hon et al. 2012).

CSPs, reflecting a general practice in the wider IT industry, attempt to minimise their liability for any loss—direct, indirect, or consequential—that may arise from the provision of the service. In cloud computing, indemnities and liabilities are usually related to privacy and security breaches and resulting data loss, data misuse and associated regulatory penalties, but may also include service interruptions or outages, or otherwise failing to meet agreed service levels (Hon and Millard 2018; Leimbach et al. 2014; Bradshaw et al. 2013). It should be noted that CSPs, typically attempt to compensate, where possible, for underperformance through service credits. Obviously, this goes to the heart of trust, particularly where critical systems have been outsourced to a CSP. Trust literature suggests that trust repair is more effective when complemented with substantive actions including admission of fault and penance signals (Bachmann et al. 2015). However, in practice, it may be more nuanced. Where a cloud service is unavailable and business is adversely impacted, service credits for the same service are unlikely to be desirable or adequate compensation. Furthermore, CSPs will often seek to exclude a wide range of under-performance and impose limitations on how service credits can be used (Bradshaw et al. 2011).

As discussed above, CSPs may try to achieve such limitations on their liability by specifying a preferential jurisdiction in the TOS. For example, US courts have enforced such limitations on liability for click-wrap agreements (see, for example, *Treiber & Straub, Inc. v. United Parcel Service, Inc.*). Given that the research referred to above suggests that consumers are not aware of the detail of the contracts they are agreeing to, and if they were, for the most part would still proceed, authors have suggested that the US courts should rejuvenate the doctrine of unconscionability to help cloud clients avoid waiving important legal rights (Calloway 2012). Notwithstanding this, EU law provides some protection against the exclusion of liability (see *GB Gas Holding v Accenture*). For individual consumers, the EU Unfair Terms Directive (Directive 93/13/EC) requires that contracts must be drafted in such a way to prevent the imposition of terms prejudicial to consumer rights. It introduces the notion of "good faith" in order to prevent significant imbalances in the dealing of consumers and suppliers. Article 5 of the Directive requires contract terms be drafted in plain and intelligible language and states that ambiguities will be interpreted in favour of consumers. Similarly, the EU Consumer Rights Directive (Directive 2011/83/EU) highlights the requirement for suppliers to provide specific information in a "clear and comprehensible manner." It also provides formal requirements and withdrawal rights for distance contracts. In 2022, new protections for consumers will be introduced as part of the Digital Content Directive (Directive 2019/770/EU) when they purchase digital services or digital content, or particularly relevant in the case of cloud services, when they exchange personal data that goes beyond the minimum necessary to provide the service. As a final comment, in Europe, data protection is a fundamental right set out in Article 8 of the EU Charter of Fundamental Rights. Article 82 of the GDPR provides for compensation for persons suffering damage due to unlawful processing or of an act incompatible with national data protection law.

2.4.3   Acceptable Use Policies

AUPs are typically incorporated or referenced in the TOS, and are used by CSPs, nominally, to protect themselves in the event of misconduct by their client, and their clients' end users. In effect, AUPs set out a largely homogenous list of prohibited activities and behaviours and the consequences for misuse (O'Byrne 2019; Bradshaw et al. 2011). Common categories of prohibited activities include:

1. activities that engage in, foster, solicit or promote illegal, abusive or irresponsible behaviour e.g. fraud, hacking, hosting and distributing viruses, or abusive, offensive or morally repugnant content e.g. child pornography, excessive violence, hate speech etc.;
2. high risk use where the failure or fault of the cloud service could result in death or serious bodily or to physical or environmental damage e.g. use in air transportation, nuclear or chemical facilities;
3. non-consensual e-mail, advertising, tracking or other uses of personal data e.g. using cloud services to spam third parties with email or advertising; and
4. abusive or offensive behaviour towards a member of the CSP staff.

This is not an exhaustive list, yet one can see that many of these activities could preclude perfectly legal activities, e.g. healthcare, and many involve a judgment by the CSP, the basis of which is typically unclear. AUPs are often neglected by clients yet can result in suspension or termination of end user accounts or indeed the client's overarching agreement. Furthermore, CSPs often retain the right to vary the terms of the AUP independently of the main TOS. For enterprise clients, aligning their AUP and their CSP's AUP is critical, otherwise an end user may have an account terminated by the CSP while the enterprise client is still accountable for delivering the service (Hon et al. 2012). Ideally, enterprise clients should negotiate a process that may be more appropriate for their needs, e.g. that they, the enterprise client, should inform their end users of AUPs, end user account suspensions, or terminations. Hon et al. (2012) note that such negotiations would seem to be the exception rather than the rule.

2.4.4   Data Protection and Privacy Policies

Issues relating to data protection and privacy can be found in the TOS, SLA, AUP and, of course, the privacy policy. It is worth noting that the privacy policy often primarily relates to CSP collection and use of personally-identifiable data. Chapters 3, 4, and 5 discuss data protection and privacy in much greater detail, however three contractual aspects are worthy of note, namely data protection, data integrity and data availability.

CSPs are required to comply with data protection regulations. Under the GDPR, CSPs are typically "data processors" but may be "data controllers" in their own right; similarly, CSP clients are typically data controllers. Article 32 of the GDPR requires the data controller and the data processor "to implement appropriate technical and organisational measures to ensure a level of security appropriate to the risk…[including] measures to protect data from accidental or unlawful destruction, loss, alteration, unauthorised disclosure of, or access to personal data transmitted, stored or otherwise processed". Article 44 deals with transfers outside the EU and only allows such data transfer subject to the GDPR. Under the GDPR, the contract between a CSP and their client must stipulate that the data processor will only act on the instructions from the data controller. One area of potential friction is that of security. The extent to which a client can instruct a CSP on their security policies in a multi-tenant commoditized infrastructure is limited and CSPs have relied on adherence to industry certifications or best practice frameworks to overcome client and regulator concerns e.g. PCI-DSS, ISO27001, COBIT etc. At the same time, CSPs, typically reserve the right to change their security policies unilaterally (Leimbach et al. 2014). While such certifications are envisaged by the GDPR under Article 42, they are not obligatory. Being 'certified' does not equate to GDPR compliance; it merely certifies that the aforementioned technical and organizational measures are in place. Indeed, the issue of certification would seem to be an area still couched in ambiguity. The European Data Protection Board only issued guidelines on GDPR certification in June 2019 and it is unclear whether certification commonly cited by CSPs meets these guidelines at the time of writing (EDPB 2019).

Data integrity is often referred to but poorly defined. For example, there are ambiguities even between information and data integrity (Boritz 2005). Notwithstanding this, it is widely accepted that it is synonymous with representational faithfulness. In contrast, data availability is the extent to which an organisation's full set of computational resources are accessible and usable (Jansen and Grance 2011). Extant pre-GDPR research suggests that, at least historically, CSPs attempted to place responsibility for preserving data integrity and backup with the client (Bradshaw et al. 2011; Hon et al. 2012). Article 32 (1) of the GDPR requires data controllers and data processors to have the ability (1) to ensure the ongoing confidentiality, integrity, availability and resilience of processing systems and services; and (2) to restore the availability and access to personal data in a timely manner in the event of a physical or technical incident. While the GDPR applies to personal data, this does not guarantee the integrity and availability of all data, for example non-personal business data, and only applies to data within the definition of the GDPR. As such, care should be taken by organisations seeking to outsource operations to the cloud, particularly where one of the motivations is that the cloud is a safe way to back up data.

2.4.5   *Variation in Terms*

As referenced earlier, CSPs typically reserve the right to change contract terms and policies unilaterally. Such variation may be communicated by reference to an updated version of the TOS, the AUP, the SLA and the privacy policy on the CSP's website. This is particularly the case in consumer and free cloud services and can result in changes to the specific services being consumed or the service levels (Michels et al. 2019; Kamarinou et al. 2015; Hon et al. 2012; Bradshaw et al. 2011). In many of these cases, the only option for clients and end users is to take it or leave it. Clients may or may not be notified of changes.

2.4.6   *Intellectual Property*

A number of issues arise in relation to intellectual property (IP) rights that should be addressed in cloud computing contracts. Spulber (2018) posits that current contracts based on tangible services are not suitable for modern technological paradigms, such as cloud computing, as they neither fully recognise the complete spectrum of IP, exclusion of access, and transferability of non-rivalrous intangible assets, nor do they address problems that arise from intentional or unintentional cooperative contribution to the creation of intangible assets. Cloud computing raises significant issues in relation to the four main categories of IP, namely—trade secrets, patents, trademarks and copyright.

The complexity of the chain of service provision in cloud computing complicates IP management. In addition to the primary client outsourcing systems to the CSP, a wider number of stakeholders may be involved in the transport, processing and storage of data, many of which may not be privy to the initial agreement with the client. Excluding access to this data while meeting SLAs may not be feasible. This may result in inadvertent disclosure of trade secrets and confidential information generally and result in civil and criminal liabilities. In the case of patents, the distribution of confidential information relating to a proposed invention may constitute a form of public knowledge of prior art and can invalidate a patent. Given the opaqueness of the chain of service provision in cloud computing, such an infringement may be difficult to prove. CSPs make use of a wide range of proprietary, third party and open source software in the delivery of their services, and will often attempt to exclude warranties on IP relating to such software, and particularly open source software (Hon et al. 2012). At the same time, AUPs will often include infringement of IP as a prohibited activity. Again, software indemnities tend to be one-sided in cloud contracts favouring the CSP.

Despite persistent rumours that social networking sites and other CSPs are attempting to claim rights in images loaded on to their systems, recent research suggests that CSPs do not seek to have copyright assigned to them but in many cases explicitly acknowledge that the end user retained the copyright (Michels et al. 2019). The ownership of metadata is less clear. Metadata is data about data and is often a by-product generated from the interaction of the clients and their end users with the cloud service. In this way, new data (which may be of value and therefore be an intangible asset) is created by the cooperation of the client, or their end users, and the CSP. While some data is used for cloud service optimisation, other data may be collected with no specific purpose in mind. This data, sometimes referred to as 'exhaust data' or 'digital data exhaust', may have significant value to third parties through data mining, aggregation or other data analytics techniques. Reed (2010) suggests that data generated by the CSP for its own internal purposes belongs to them, however if the data contains client data protected under copyright, the client may have an infringement claim—if the client is aware of such use at all. Reed (2010) suggests CSPs need to pay careful attention that they do not take unfair advantage of clients nor infringe copyrighted works. But what of digital data exhaust? Who owns this data? CSPs are typically silent on this. Indeed, it may be a case of 'don't ask, don't tell'. Nonetheless, contracts should state clearly whether such data is being collected and for what use.

Hon et al. (2012) identify similar issues relating to the ownership of software applications developed by clients or end users on a CSP's IaaS or PaaS platform where the CSPs integration tools are used or the software is designed for specific use only with that CSP's software, and is therefore tied to the CSP's IP. The emergence of cloud service brokerage (CSB) models, and in particular consumer app marketplaces (e.g. Google Play and Apple AppStore), B2B cloud application and API marketplaces (e.g. Salesforce AppExchange and RapidAPI), and indeed Marketplace as a Service models (Paulsson et al. 2016; Paulsson et al. 2020), complicate these matters further. In these cases, independent software vendors build their businesses with near-total dependency on a small number of large cloud ecosystems where the underlying CSP holds disproportionate bargaining power and data. The degree of trust involved is near total. Similarly, where clients or end users suggest or actually implement improvements or bug fixes, it may not be clear where IP ownership lies (Leimbach et al. 2014).

2.4.7   Termination

Contracts may come to a natural and expected conclusion, or be unnaturally terminated due to breach of contract or some other event rendering them invalid. On termination, the contract should make adequate provision for the consequent handling of the client data including defining the term of service and (non-) renewal of service; termination events; data preservation; data deletion; and data transfer, following termination (Leimbach et al. 2014). The treatment of termination has legal and economic implications. CSPs can use data preservation, in particular, as a means of vendor lock-in by making data transfer to another service time-consuming or cumbersome. In the event of unnatural termination, clients will want to ensure that they have adequate time to access their data and transfer their data from the incumbent CSP to an alternative. At the end of a service contract, CSPs may (1) immediately delete the data, (2) provide a grace period, or (3) offer a hybrid approach neither obliging the deletion nor preservation of data, nor undertaking to delete data and offering a grace period at their discretion (Bradshaw et al. 2011). In reality, few CSPs delete the data on termination. However against the backdrop of data protection legislation, deleting personal data as soon as possible may be prudent to mitigate GDPR-related risk. At the same time, Hon and Millard (2018) have suggested that regulators are seeking maximum periods after which personal data must be deleted.

Leimbach et al. (2014) also note that clients should understand what happens to metadata relating to their account on termination; this issue would seem to be contentious not only from a termination perspective but also from an ownership perspective and is worthy of attention by cloud clients. As discussed above, the line between the metadata that a CSP reasonable owns and that which the client owns can be blurry and ownership uncertain. A CSP generates, stores, analyses and may replicate system and network usage, data transfer and other logs as part of the activities inherent in the delivery and future development of the service. It may not be possible or desirable for CSPs to delete this data; similarly, clients may not wish inferences about them or their activities inferred from this data.

As a final comment on termination, while CSPs do not have a duty to make off-boarding easy or free, and indeed there is a palpable difference between the quality of on-boarding tools and support and off-boarding ones, they are not the only cause of delays. Customisation of cloud services, for example Salesforce.com, can result in both vendor lock-in and data portability issues. As such, cloud clients need to be aware of how integrated and dependent they are becoming on their CSP over time and the implications on termination.

2.4.8   *Dispute Settlement*

The overwhelming majority of CSPs include provisions for dispute resolution in their TOS however specific dispute resolution clauses may feature in other documents e.g. privacy policies (Martic 2017). CSPs may stipulate courts or arbitration to settle disputes and stipulate a choice of law in one or more jurisdictions (for example, see *Ryanair dac v SC Vola.ro srl*) or specific arbitration rules e.g. AAA or ICC rules. Research suggests that the preference is for courts as the exclusive adjudicative method (Martic 2017). As discussed previously, the determination of choice of law can favour one side or the other. For example, EU consumers can avail of the EU Alternative Dispute Resolution (ADR) Directive and the EU-wide Online Dispute Resolution (ODR) platform.

2.5   Future of Cloud Contracts

Current literature and thinking on contract law and cloud computing is based on relatively static conceptualisations of both cloud computing and contracts. The majority of legal research focuses on a conceptualisation of cloud computing from over a decade ago, primarily focussing on IaaS and SaaS services, and to a much lesser extent, Platform as a Service (PaaS). Recent work has suggested that the cloud is increasingly more abstracted, heterogeneous, composable, and automated (Lynn et al. 2020). First, the emergence of containerisation and serverless computing (including Function-as-a-Service) are enabling portability and a separation of concerns between CSPs and independent software vendors and clients (Lynn et al. 2020). These paradigms reduce vendor lock-in and create clear lines of demarcations between responsibilities and technology ownership in ways not envisaged by traditional cloud computing. Similarly, cloud computing is becoming more heterogeneous and composable with a wider variety of customisable configurations available to clients that impact performance and complicate service level expectations, thus pushing performance-related decisions to the client, and requiring more nuanced agreements. Furthermore, with the advent of the Internet of Things, the cloud is becoming more decentralised and distributed across a cloud-to-thing (C2T) continuum. This has resulted in new computing paradigms including fog, mist and edge computing (Iorga et al. 2018). Processing and storage may take place in the cloud, at the edge or somewhere in between (the fog).

This new decentralised, abstract, heterogeneous, and composable cloud introduces complexity at several orders of magnitude higher than today. It is beyond human capabilities to manage such infrastructure manually. As a result, the cloud is becoming even more automated and intelligent. Artificial Intelligence for IT Operations (AIOps) algorithms and machine learning monitor, operate, and maintain distributed systems (Cordoso 2019). The emergence of self-organising and self-learning systems represents a significant evolution in cloud infrastructure decision-making. Outsourcing decision-making to AI provides substantial technical, legal and trust challenges, not least the black box nature of AI decision making. It is foreseeable that the actions of AI will result in cloud under-performance at some time in the future and addressed in accordance with existing legal provisions. However commentators have noted that AI may not be recognised as a subject of law, and as a result, may not be held personally liable for the damage it causes (Čerka et al. 2015). Cloud contracts need to evolve to reflect this changing and more nuanced cloud.

At the same time, the nature of contracts is developing, albeit at a much slower pace. Spulber (2018) has proposed a new framework for 'intellectual contracts', a form of "...agreement to create, develop, share, or apply intangible assets involved in technological change." In his conceptualisation, Spulber attempts to overcome the shortcomings of traditional contracts with respect to the completeness, excludability, and transferability of intangible assets while recognising that IP arises from intentional and unintentional cooperation, and rights in such outputs needs to be addressed in contracts. Similarly, there has been renewed discussions on the value of smart contracts in cloud computing with the emergence and hype around Blockchain. Smart contracts are not new; in effect they are agreements whose execution is automated and self-enforceable. Vending machines are cited as common examples. In the mid-nineties, Szabo (1996) envisioned a computerised transaction protocol that implements the terms of a contract. While Szabo (1996) foresaw self-enforcing contracts based on conditions being met, Blockchain addressed a number of issues in smart contracts, and contracts more generally, not least the verifiability of conditions and performance. In the language of trust, Blockchain verifies integrity. Blockchain has been proposed as a solution to a number of cloud-related contract issues including verifiability of performance (Dong et al. 2017), GDPR compliance (Corrales et al. 2019), and digital rights management (Finck and Moscon 2019). Despite the benefits of smart contracts, significant questions remain unanswered regarding their enforceability (Savelyev 2017). Indeed, the extent to which either smart contracts or intellectual contracts can be easily adapted and integrated in to current contract law frameworks, or indeed need to, is open to debate.

2.6   Conclusion

This chapter provides an overview of some of the key contract law issues that arise in cloud computing. It is not exhaustive. It is clear that contracts in cloud computing are used primarily as an instrument of control and, to a lesser extent, coordination. While larger commercial and governmental organisations, may make rational choices based on calculus-based trust in full knowledge of the contract they are entering in to, it is clear that for the vast majority of firms do not have the opportunity or bargaining power to negotiate with cloud service providers. A rational, albeit disadvantageous decision, to either 'take it or leave it', remains. Given the homogeneous nature of cloud computing terms and conditions and the dominance of a small number of hyperscale players in the public cloud market, organisations, and particularly smaller ones, are left on the poorer side of a one-sided power relationship. While CSPs may possess undeniable competence, this imbalanced relationship does not foster trust. The evidence of the terms and conditions reviewed in this chapter suggest CSPs are not benevolent and their integrity can only be judged on their *post hoc* performance. Based on their approaches to limitation of liability, warranties, compensation, amongst others, one could understand how firms might tend towards scepticism. Such scepticism or lack of trust need not be a negative, it may be constructive resulting in enterprise customers developing healthy vigilance behaviours e.g. increasing monitoring, ensuring compliance, and preventing potential exploitation (Lumineau, 2017).

In contrast, individual consumers, by and large, do not make rational decisions with regards to entering in to contracts with cloud service providers. Research suggests that they do not read the terms and conditions in advance of using cloud services, and if they do, this does not change their decision. In this way, their decisions may be non-calculative and habitual, or reflect a perceived lack of options and/or a desire to avoid anxiety. At the same time, this is not their problem in one sense, but a regulatory one. There has been much progress to correct imbalances in contractual terms in consumer cloud computing, not least the GDPR. Unfortunately, regulatory responses are not uniform worldwide. Borderless technologies such as cloud computing provide significant challenges particularly in a highly globalised world. The rejuvenation of the doctrine of unconscionability as proposed by Calloway (2012) may be worthy of consideration, particularly in the US, the choice of law for so many cloud service providers.

Evolutions in cloud computing are overcoming the sources of legal friction between demand and supply. Containerisation and serverless computing introduce a separate of concerns that institutionalise a trust compartmentalisation of sorts in line with Lumineau (2017). At the same time, heterogeneity, AIOps, and the Internet of Things, further complicate the relationship between supply and demand, accountability and assurance, and as a result, trust and distrust. Innovations in contracts, whether intellectual or smart, remain at an early stage of conceptualisation. Indeed, it is unclear whether they are enforceable and can be adapted in to a legal infrastructure designed around traditional notions of tangible intellectual property and, even so, whether lawmakers will consider it necessary at all.

2.7   CASES

Caspi v. Microsoft Network LLC, 732 A.2d 528 (N.J. Super. 1999)
Data Protection Commissioner v Facebook Ireland Limited, Maximilian Schrems (Case C-311/18)
El Majdoub v CarsOnTheWeb.Deutschland GmbH Case [2015] EUECJ C-322/14
Gas Holdings Limited v Accenture (UK) Limited and others [2010] EWCA Civ 912
Rudder v Microsoft Corp [1999] OJ No 3778 (Sup Ct J)
Ryanair dac v SC Vola.ro srl [2019] IEHC 239
Treiber & Straub, Inc. v. United Parcel Serv., Inc., No. 04-C-0069, 2005 WL2108081 (E.D. Wis. Aug. 31, 2005).

REFERENCES

Bachmann, R., Gillespie, N., & Priem, R. (2015). Repairing Trust in Organizations and Institutions: Toward a Conceptual Framework. *Organization Studies, 36*(9), 1123–1142.

Bakos, Y., Marotta-Wurgler, F., & Trossen, D. R. (2014). Does Anyone Read the Fine Print? Consumer Attention to Standard-Form Contracts. *The Journal of Legal Studies, 43*(1), 1–35.

Bellia, A. J., Jr. (2002). Promises, Trust, and Contract Law. *American Journal of Jurisprudence, 47*, 25.

Boritz, J. E. (2005). IS Practitioners' Views on Core Concepts of Information Integrity. *International Journal of Accounting Information Systems, 6*(4), 260–279.

Bradshaw, S., Millard, C., & Walden, I. (2011). Contracts for Clouds: Comparison and Analysis of the Terms and Conditions of Cloud Computing Services. *International Journal of Law and Information Technology, 19*(3), 187–223.

Bradshaw, S., Millard, C., & Walden, I. (2013). Standard Contracts for Cloud Services. In *Cloud Computing Law* (pp. 39–72). Oxford: Oxford Scholarship Online.

Calloway, T. J. (2012). Cloud Computing, Click-Wrap Agreements, and Limitation on Liability Clauses: A Perfect Storm. *Duke Law & Technology Review, 11*, 163.

Čerka, P., Grigienė, J., & Sirbikytė, G. (2015). Liability for Damages Caused by Artificial Intelligence. *Computer Law & Security Review, 31*(3), 376–389.

CJEU (2019) - no longer required..

Cordoso, J. (2019). *The Application of Deep Learning to Intelligent Cloud Operation.* Paper presented at Huawei Planet-scale Intelligent Cloud Operations Summit, Dublin, Ireland, 1 November 2019.

Corrales, M., Jurčys, P., & Kousiouris, G. (2019). Smart Contracts and Smart Disclosure: Coding a GDPR Compliance Framework. In *Legal Tech, Smart Contracts and Blockchain* (pp. 189–220). Singapore: Springer.

Dong, C., Wang, Y., Aldweesh, A., McCorry, P., & van Moorsel, A. (2017, October). *Betrayal, Distrust, and Rationality: Smart Counter-Collusion Contracts for Verifiable Cloud Computing.* Proceedings of the 2017 ACM SIGSAC Conference on Computer and Communications Security, pp. 211–227.

EDPB. (2019). Guidelines 1/2018 on Certification and Identifying Certification Criteria in Accordance with Articles 42 and 43 of the Regulation. Retrieved from https://edpb.europa.eu/our-work-tools/our-documents/smjernice/guidelines-12018-certification-and-identifying-certification_en

Finck, M., & Moscon, V. (2019). Copyright Law on Blockchains: Between New Forms of Rights Administration and Digital Rights Management 2.0. *IIC-International Review of Intellectual Property and Competition Law, 50*(1), 77–108.

Gartner. (2019). Market Share Analysis: IaaS and IUS, Worldwide, 2018. *Gartner*.

Hon, W. K., & Millard, C. (2018). Banking in the Cloud: Part 3–Contractual Issues. *Computer Law & Security Review, 34*(3), 595–614.

Hon, W. K., Millard, C., & Walden, I. (2012). Negotiating Cloud Contracts: Looking at Clouds from Both Sides Now. *The Stanford Technology Law Review, 16*, 79.

IDC. (2013). *IDC Predictions 2013: Competing on the 3rd Platform*. IDC.

Iorga, M., Feldman, L., Barton, R., Martin, M. J., Goren, N. S., & Mahmoudi, C. (2018). *Fog Computing Conceptual Model*. (No. Special Publication (NIST SP)-500-325).

Jansen, W., & Grance, T. (2011). *Guidelines on Security and Privacy in Public Cloud Computing*. SP 800-144.

Kamarinou, D., Millard, C., & Hon, W. K. (2015). *Privacy in the Clouds: An Empirical Study of the Terms of Service and Privacy Policies of 20 Cloud Service Providers*. Queen Mary School of Law Legal Studies Research Paper, p. 209.

Kimel, D. (2001). Neutrality, Autonomy, and Freedom of Contract. *Oxford Journal of Legal Studies, 21*(3), 473–494.

Lambert, R., & Peppard, J. (2013). The Information Technology–Organizational Design Relationship. In R. D. Galliers & D. E. Leidner (Eds.), *Strategic Information Management* (pp. 427–459). Routledge.

Leimbach, T., Hallinan, D., Bachlechner, D., Weber, A., Jaglo, M., Hennen, L., Nielsen, R. O., Nentwich, M., Strauss, S., Lynn, T., & Hunt, G. (2014). *Potential and Impacts of Cloud Computing Services and Social Network Websites*. Publication of Science and Technology Options Assessment.

Lumineau, F. (2017). How Contracts Influence Trust and Distrust. *Journal of Management, 43*(5), 1553–1577.

Lynn, T., & Rosati, P. (2017). Challenges to Technology Implementation. In M. Quinn & E. Strauss (Eds.), *The Routledge Companion to Accounting Information Systems*. Routledge.

Lynn, T., Rosati, P. & Fox, G. (2020). Measuring the Business Value of Cloud Computing: Emerging Paradigms and Future Directions for Research. In T. Lynn, J. Mooney, P. Rosati & G. Fox (Eds.), *Measuring the Business Value of Cloud Computing*. Palgrave-Macmillan.

Marotta-Wurgler, F., & Chen, D. L. (2012). Does Contract Disclosure Matter? *Journal of Institutional and Theoretical Economics (JITE)/Zeitschrift für die gesamte Staatswissenschaft*, 94–123.

Martic, D. (2017). *Dispute Resolution for Cloud Services: Access to Justice and Fairness in Cloud-Based Low-Value Online Services*. (Doctoral Dissertation), Alma Mater Studiorum Università di Bologna. Dottorato di ricerca in Law, Science and Technology, 28 Ciclo.

Mayer, R. C., Davis, J. H., & Schoorman, F. D. (1995). An Integrative Model of Organizational Trust. *Academy of Management Review, 20*(3), 709–734.

McKnight, D. H., Choudhury, V., & Kacmar, C. (2002). Developing and Validating Trust Measures for e-commerce: An Integrative Typology. *Information Systems Research, 13*(3), 334–359.

Mell, P., & Grance, T. (2011). *The NIST Definition of Cloud Computing.* Gaithersburg, MD: National Institute of Standards and Technology.

Michels, J. D., Millard, C., & Joshi, S. (2019). *Beyond the Clouds, Part 1: What Cloud Contracts Say About Who Owns and Can Access Your Content.* Queen Mary School of Law Legal Studies Research Paper, p. 315.

O'Byrne, W. I. (2019). Acceptable Use Policies. *The International Encyclopedia of Media Literacy*, 1–6.

Paulsson, V., Emeakaroha, V., Morrison, J., & Lynn, T. (2016). Cloud Service Brokerage: A systematic Literature Review Using a Software Development Lifecycle. In 22nd Americas Conference on Information Systems, AMCIS 2016, CA, USA: San Diego.

Paulsson, V., Emeakaroha, V., Morrison, J., & Lynn, T. (2020). Cloud Service Brokerage: Exploring Characteristics and Benefits of B2B Cloud Marketplaces. In T. Lynn, J. Mooney, P. Rosati, & G. Fox (Eds.), *Measuring the Business Value of Cloud Computing.* Palgrave Macmillan.

Reed, C. (2010). *Information 'Ownership' in the Cloud.* Queen Mary School of Law Legal Studies Research Paper, p. 45.

Rousseau, D. M., Sitkin, S. B., Burt, R. S., & Camerer, C. (1998). Not So Different after All: A Cross-Discipline View of Trust. *Academy of Management Review, 23*(3), 393–404.

Savelyev, A. (2017). Contract law 2.0:'Smart' Contracts as the Beginning of the End of Classic Contract Law. *Information & Communications Technology Law, 26*(2), 116–134.

Schwartz, A., & Wilde, L. L. (1978). Intervening in Markets on the Basis of Imperfect Information: A Legal and Economic Analysis. *The University of Pennsylvania Law Review, 127*, 630.

Spulber, D. F. (2018). Intellectual Contract and Intellectual Law. *Journal of Technology Law & Policy, 23*, 1.

Szabo, N. (1996). Smart Contracts: Building Blocks for Digital Markets. *EXTROPY: The Journal of Transhumanist Thought, 16*, 18.

van der Wees, A., Daniele, C., Jesus, L., Edwards, M., Schifano, N., & Maddalena, S. L. (2014). *Cloud Service Level Agreement Standardisation Guidelines.* C-Sig SLA, pp. 1–41.

van der Werff, L., Legood, A., Buckley, F., Weibel, A., & de Cremer, D. (2019). Trust Motivation: The Self-Regulatory Processes Underlying Trust Decisions. *Organizational Psychology Review, 9*(2–3), 99–123.

Weber, J. M., Malhotra, D., & Murnighan, J. K. (2004). Normal Acts of Irrational Trust: Motivated Attributions and the Trust Development Process. *Research in Organizational Behavior, 26*, 75–101.

Weber, R. H., & Staiger, D. N. (2014). Cloud Computing: A Cluster of Complex Liability Issues. *European Journal of Current Legal Issues, 20*(1), 1–13.

CHAPTER 3

Competing Jurisdictions: Data Privacy Across the Borders

Edoardo Celeste and Federico Fabbrini

Abstract Borderless cloud computing technologies are exacerbating tensions between European and other existing regulatory models for data privacy. On the one hand, in the European Union (EU), a series of data localisation initiatives are emerging with the objective of preserving Europe's digital sovereignty, guaranteeing the respect of EU fundamental rights and preventing foreign law enforcement and intelligence agencies from accessing personal data. On the other hand, foreign countries are unilaterally adopting legislation requiring national corporations to disclose data stored in Europe, in this way bypassing jurisdictional boundaries grounded on physical data location. The chapter investigates this twofold dynamic by focusing particularly on the current friction between the EU data protection approach and the data privacy model of the United States (US) in the field of cloud computing.

E. Celeste (✉) • F. Fabbrini
School of Law and Government, Dublin City University, Dublin, Ireland
e-mail: edoardo.celeste@dcu.ie; federico.fabbrini@dcu.ie

T. Lynn et al. (eds.), *Data Privacy and Trust in Cloud Computing*,
Palgrave Studies in Digital Business & Enabling Technologies,
https://doi.org/10.1007/978-3-030-54660-1_3

43

Keywords Cloud computing • Data privacy • Data protection • Data localisation • Data residency • Digital sovereignty

3.1   Introduction

Over the past decade, the right to privacy and the protection of personal data have been increasingly recognised as fundamental values at global level (Greenleaf 2019; Bygrave 2014; Solove 2008). Yet, their understanding still varies significantly among jurisdictions. One apparent example is offered by the different approach to data privacy in the European Union (EU) and the United States (US). In Europe, data protection is a constitutionalised fundamental right, and a comprehensive set of legislation has been put in place to make the regulation of personal data processing uniform across member states. Conversely, in the US, data privacy is not explicitly enshrined in the federal constitution, and is regulated only in selected pieces of legislation targeting specific sectors considered worthy of intervention.

Divergence in legal frameworks of data protection is certainly not a novelty of the last decade. However, the recent development of borderless digital technologies, such as cloud computing, amplifies the risk of tensions between different regulatory models. When data are stored in the cloud, it becomes more difficult to identify the applicable law easily. In response to this phenomenon, data localisation initiatives requiring data to be physically stored in servers located within national boundaries have recently emerged as a regulatory trend to avoid conflicts of law, enhance the level of data privacy protection, limit the risk of access from foreign intelligence agencies, and facilitate domestic law enforcement.

This chapter investigates this twofold dynamic by focusing on the current friction between the EU data protection approach and the US data privacy model in the context of cloud computing. The chapter is structured as follows. In Sect. 4.2, we discuss the main areas of divergence between EU and US approach to data privacy. Then, in Sect. 4.3 we explain how these differences create a series of regulatory challenges in the context of cloud computing. Section 4.4 analyses how recent legal and policy developments on both sides of the Atlantic are addressing these issues, with a particular focus on data localisation initiatives and strategies to preserve digital sovereignty. The chapter concludes with the proposition that data localisation does not represent a panacea for resolving tensions between competing jurisdictions in the field of cloud computing, and that transnational cooperation and effective international agreements are needed now, more than ever.

3.2   DATA PRIVACY ACROSS THE ATLANTIC

In Europe and the US, data privacy law emerged almost simultaneously in the 1970s (Jones 2017). Both legal systems recognise the importance of protecting personal data and the potential risks deriving from a misuse of such data. Yet, on the two sides of the Atlantic, two different regulatory models have emerged in the field of data privacy (Schwartz and Solove 2014; Tourkochoriti 2014).

In Europe, the respect of privacy and the protection of personal data are recognised as fundamental rights. In 1950, as a reaction to intrusive surveillance practices of totalitarian regimes that afflicted Europe in the first half of the twentieth century, the European Convention on Human Rights enshrined the individual right of respect for private and family life, home and correspondence (Article 8). In its case law, the European Court of Human Rights, which is the competent jurisdiction for the interpretation of the Convention, has affirmed that the concept of private life must be construed broadly in order to protect all aspects of human personality, including individual personal data (Council of Europe 2019; Fabbrini 2015). In 2000, the EU Charter of Fundamental Rights explicitly enshrined the right to privacy and data protection in two distinct provisions, Articles 7 and 8, respectively. Although originally lacking binding legal value with the transposition of the Lisbon Treaty in 2009, the Charter was recognised as having a primary legal status in the hierarchy of EU legal sources, at the same level of EU founding treaties (Fabbrini 2015).

The US is often referred to as the cradle of the right to privacy. Back in 1890, Samuel Warren and Louis Brandeis authored a seminal article published on the Harvard Law Review in which they advocated for the recognition of a broad conceptualisation of the right to privacy and the protection of the individual against external intrusions (Warren and Brandeis 1890). However, in contrast to the EU, in the US, at least at federal level, there is no explicit constitutional provision protecting the right to privacy or data protection. Indeed, the US Constitution dates to 1787 and its Bill of Rights was added only three years later, so well before privacy became an issue. The case law of the US Supreme Court progressively recognised different aspects of privacy, regarded both as a negative right against State intrusion and as a positive right to self-determination in a variety of contexts, including the choice of using contraceptives or terminating pregnancy (Flaherty 1991). Lacking an explicit reference, the US Supreme Court had to find a constitutional support for the right to privacy in the "emanations" and "penumbras" of the Bill of Rights (*Griswold v. Connecticut*, 381 U.S. 484). In particular, they examined the Fourth Amendment, protecting citizens against unreasonable search and seizures (Solove 2001), and the Fourteenth Amendment, subjecting any deprivation of life, liberty and property to due process rules (Cate and Cate 2012).

Besides the different constitutional frameworks, the EU and the US also developed alternative regulatory models in the field of data privacy. Over the past few decades, the EU has introduced a fully comprehensive set of legislation governing the processing of personal data, both in the private and in the public sector (Fabbrini 2015). In 2016, the EU replaced the 1995 Data Protection Directive, which represented the core piece of legislation adopted to harmonise national statutes in the field, with a General Data Protection Regulation (GDPR), whose provisions are directly binding in all member states (Albrecht 2016). Conversely, the US have rejected a similar all-encompassing approach, in favour of exclusively regulating specific sectors which were felt to be more in need of intervention (Schwartz and Solove 2014). Although being a pioneer in the data privacy field, having adopted the Privacy Act 1974, which regulates data processing by federal agencies, the US never introduced a unitary and comprehensive piece of legislation in the field of data privacy, and only few US states have. At the federal level, US data privacy law is a mosaic of normative instruments covering a variety of issues, spanning from children's privacy to the use of data in financial services (Schwartz and Solove 2014).

In Europe, the basic presumption is that processing personal data represents an interference with the right to data privacy that can be tolerated only if it satisfies certain legal conditions. In the US, instead, data processing is considered fully legitimate in so far as it is not prohibited by law, and a strong emphasis is placed on the role of individual consent as a basis to process personal data (Tourkochoriti 2014). European data protection law, in order to reduce the risk of circumvention and ensure an even level of protection across member states, has introduced provisions extending its application to data controllers that are not established in the EU, but nevertheless process data related to EU residents (Article 3 GDPR; Christopher Kuner 2015; Svantesson 2015; de Hert and Czerniawski 2016). In the US, data privacy statutes do not have a similar extraterritorial effect.

Lastly, in contrast to US legislation, EU data protection law also regulates international data transfers. Article 44 GDPR establishes that personal data can freely circulate among member states, but cannot be transferred to third countries unless they provide an adequate level of protection. Article 48 of the GDPR even explicitly prohibits any data disclosure requested by a foreign authority, unless based on an international treaty. The European Commission can adopt a decision certifying the adequacy of the level of data protection of a third country (Article 45 GDPR). Countries like Israel, Argentina, Uruguay, and recently Japan, have been certified as providing an adequate level of protection (European Commission 2019). Conversely, the Commission has only issued a partial adequacy decision in relation to the United States.

In 2000, the European Commission adopted Decision 2000/520/EC (so called "Safe Harbor") which established the adequacy of US data protection rules: in particular, US corporations that are subject to the supervision of the Federal Trade Commission could self-certify their respect of the Safe Harbor Principles (Greer 2011). However, in the aftermath of the Snowden revelations about the existence of US mass surveillance programmes, this decision was invalidated by the European Court of Justice (ECJ). In the *Schrems* case (C-362/14), decided in 2015, the ECJ held that the Commission, by certifying the adequacy of the Safe Harbor scheme, failed to take into account the power of US law enforcement authorities to access on a generalised basis EU data transferred under the Safe Harbor scheme (Cole and Fabbrini 2016; Padova 2016). According to the ECJ, such a model of bulk surveillance cannot be tolerated as it compromises the essence of the right to privacy protected by the EU Charter of Fundamental Rights (para 94 of the judgment; see Ojanen 2017).

The Safe Harbor scheme was promptly replaced by the so-called "Privacy Shield", which was negotiated between the European Commission and the US authorities in 2016 and entered into force with Decision 2016/1250. The new system is very similar to the Safe Harbor in terms of functioning, but has been accompanied by a series of further guarantees, especially in relation to the individual right of redress (Tracol 2016; cf. Bender 2016). Moreover, after the Snowden revelations, the US started a progressive revision of its law enforcement legislation (Cole and Fabbrini

2016). Nevertheless, recently, a new legal challenge was made against the EU-US Privacy Shield and in July 2020 the ECJ declared also this instrument invalid for breach of EU data privacy law. (Case C-311/18, *Data Protection Commissioner v Facebook Ireland Limited, Maximilian Schrems*). In its ruling, the ECJ emphasized the same problem as in *Schrems I*: the level of protection of EU data in the US is still contested.

The *Schrems I* and *Schrems II* cases both represent examples of circumstances in which the EU and US data protection frameworks enter in conflict. This situation arises when transnational processing of data is involved, and is highly problematic both from a EU and US perspective. On the one hand, EU data protection law imposes limits to the free transfer of personal data to third countries that are not deemed to offer an adequate level of protection of personal data. On the other hand, US authorities are loath of bending their sovereign decisions to EU requests in the field of data privacy as a result of the so-called Brussels effect (Bradford 2012). As the next section will explain, cloud computing, by ordinarily involving transborder data processing, represents a particularly challenging area.

3.3   REGULATING BORDERLESS CLOUD COMPUTING

Cloud computing denotes "flexible, location-independent access to computing resources that are quickly and seamlessly allocated or released in response to demand" (Hon et al. 2011a, p. 6). This broad definition encompasses three models of cloud computing: Infrastructure as a Service (IaaS), Platform as a Service (PaaS), and Software as a Service (SaaS). These models, as is apparent from their denomination, differ on the basis of the service offered, spanning from the mere provision of infrastructure to the supply of software (Hon et al. 2011a). These paradigms, however, are not mutually exclusive. It is conversely possible that a cloud computing service is composed of infrastructure, platform or service layers at the same time (Hon et al. 2011a). Just to mention some familiar examples in the academic context, Dropbox, the Google apps and Microsoft 365 represent commonly used Software as a Service cloud computing services.

A further classification of cloud computing models takes into account their users: one can distinguish between public, private or hybrid cloud computing models (Esayas 2012; see also Varadi et al. 2012). In the first case, cloud computing services are available to the general public, an example being the social network Facebook; in the second case, their use is restricted to a limited number of users, such as in tailored cloud services for corporations or institutions; lastly, the third case represents an intermediary solution.

Using computing resources which are available in "the cloud" is advantageous for a series of reasons (Hon et al. 2011a; Esayas 2012). First of all, cloud computing can provide services which are tailored to the end user. Secondly, cloud computing can flexibly respond to changes in users' demand. And lastly, but certainly not least, cloud computing is significantly cheaper than developing and maintaining individually owned infrastructure, platforms or software. Those resources are centralised, and thanks to their virtual character, they are shared according to the specific needs of potential users.

From a technical perspective, this is possible thanks to the so-called "sharding" (Hon et al. 2011a). Data are not concentrated in a single virtual cloud, but are fragmented into a series of "shards", replicated, and stored in different locations. This procedure, which is entirely automated, allows the cloud computing service to maximise its performance. On the one hand, smaller pieces of information can be accessed more quickly. On the other hand, their replication enhances the security of the system by reducing the risks of node failures or data loss.

The technical architecture of cloud computing creates a series of challenges from a data protection perspective. First of all, cloud computing providers may be unaware of the fact that they are processing personal data. Hon et al. talk of the "cloud of unknowing" (2011a, p. 1). Secondly, the multi-layered structure of cloud computing services may create issues in relation to the correct identification of the data controller and processor, and the consequent allocation of responsibilities. For example, it has been contended that cloud service providers merely offering infrastructure as a service can even hardly be considered as data processors (Hon et al. 2011b). Thirdly, cloud computing models may involve a continuous transfer of data on a global scale, and therefore potentially interesting a multiplicity of states. The "sharding" procedure, on which cloud computing relies, partitions and transfers data automatically.

The introduction of the GDPR has removed a series of jurisdictional problems existing under the Data Protection Directive. The GDPR is immediately legally binding in all EU member states. As a consequence, at least if the transfer occurs within the EU, the data controller will have one single legislative reference point instead of multiple different domestic pieces of legislation. Moreover, the GDPR has eliminated the reference to the use of equipment situated in an EU member state as a criterion to define its scope of application. The idea of linking the applicability of EU data protection law to the physical use of an equipment no longer corresponded to the technological reality (Hon et al. 2012; Esayas 2012; Christopher Kuner 2010). The GDPR now regulates data controllers who are not established in the EU, but offer goods or services in the EU or monitor the behaviour of European data subjects (Article 3 GDPR). However, the GDPR has not substantially modified the data transfer regime involving third countries. Therefore, data controllers should still ensure that, when using cloud computing services, European data are not transferred to third countries which do not guarantee an adequate level of protection, or without appropriate safeguards (Hon and Millard 2012).

The existence of these regulatory obstacles to the free flow of personal data from the EU to third countries has led cloud computing providers to offer services storing personal data on servers exclusively located in the EU (Hon and Millard 2012). EU data protection law has been one of the main drivers behind the creation of "regional" clouds besides cross-border ones (Svantesson and Clarke 2010). A tension therefore emerges between, on the one hand, the economic and technological dimensions that push towards the offer of cloud computing services on a global scale in order to maximise efficiency and minimise costs, and, on the other hand, regulatory and policy initiatives that conversely impose boundaries and *de facto* limit the free flow of data for privacy rights reasons. Since the main cloud computing providers are based in the US and, as pointed out above, the EU and US are adopting different approaches in relation to data privacy, this situation raises several challenges. The next section will examine a series of initiatives that are emerging on both sides of the Atlantic to address these problems.

3.4   DATA LOCALISATION AND DIGITAL SOVEREIGNTY

Over the past few years, data localisation—which is the requirement to store data in servers located within a given jurisdiction—has also emerged as a regulatory trend at global level (Mishra 2015; Selby 2017). To mention a successful example, in 2014 Russia introduced a statute requiring citizens' personal data to be stored in the national territory (Hon et al. 2016; Selby 2017). The objectives of these kinds of legislation are disparate. Safeguarding data privacy and ensuring effective law enforcement at domestic level are the two most recurrent explicit justifications of these initiatives (Mishra 2015; Hon et al. 2016). The timing of this phenomenon, which has thrived after the Snowden revelations about the US mass surveillance programmes, also suggests that data localisation is emerging as a response to the risk of data access from foreign intelligence agencies (Hon et al. 2016).

In Europe too, a series of data localisation initiatives has recently emerged. Since 2011, ideas of a Europe-only cloud, if not even a "virtual Schengen area", have been circulating (Kuner et al. 2015; Hon et al. 2016). In 2013, the German telecommunications operator, Deutsche Telekom announced a plan to create a German "Internetz", by ensuring that traffic data are only routed nationally (Hon et al. 2016). Similarly, after Russia's annexation of Crimea in 2014, Estonia explored the possibility of creating a "data embassy" via a combination of a physical diplomatic seat in a friend country to locate data centres, and a "virtual embassy" in a private cloud to store critical data (Millard 2015).

More recently, the European Commission has launched a European Cloud Initiative in the context of its Digital Single Market Strategy (European Commission 2016). This policy includes the creation of a European Open Science Cloud, which aims to offer European researchers a safe environment to store and share data, and a European Data Infrastructure, which would provide the necessary super-computing solutions. Moreover, in 2019, the German Ministry for Economic Affairs and Energy has officially presented 'Gaia-X', the project for a European federated cloud-based data infrastructure (Federal Ministry for Economic Affairs and Energy (BMWi) 2019).

These initiatives show that the concept of "digital sovereignty" has recently emerged as a common thread in the European debate on data localisation. Originally, proposals such as the virtual Schengen area were politically justified by the need to ensure a sufficient level of security in the digital environment (Hon et al. 2016). The protection of human rights, and in particular the rights to privacy and data protection, has been the second main driver of discussions about data localisation in Europe. In the *Digital Rights Ireland* case, for example, the ECJ invalidated Directive 2006/24/EC, compelling telecommunications operators to retain all users' metadata for a fixed period of time, on the basis, *inter alia*, that it failed to require the storage of personal data in Europe (Digital Rights Ireland 2014, para. 68; Celeste 2019). According to the ECJ, the Data Retention Directive, by allowing telecommunications operators to store retained meta-data outside Europe, undermined the power of member states' national data protection authorities to control data processing, as expressly prescribed by Article 8(3) of the EU Charter of Fundamental of Rights (Digital Rights Ireland 2014, para. 68; cf. Tele2 Sverige 2016, para. 122).

More recently, in the summer 2019, the data protection authority of the German Land of Hessen temporarily ordered Hessian schools not to use Microsoft Office 365 (Der Hessische Beauftragte für Datenschutz und Informationsfreiheit 2019a; cf. Walden 2011). The decision followed Microsoft's announcement that the company would not ensure data storage on the German cloud only. The supervisory authority found that the risk of allowing US authorities to access European children's data without appropriate guarantees made the use of Microsoft's software unacceptable from a fundamental rights perspective (Der Hessische Beauftragte für Datenschutz und Informationsfreiheit 2019a, para. 2). The Hessian ban, which was originally extended to Google and Apple cloud applications (Der Hessische Beauftragte für Datenschutz und Informationsfreiheit 2019a, para. 5), was subsequently lifted a month later following an intense phase of dialogue with Microsoft. The supervisory authority, however, stated that the investigation would have continued in light of several legal and technical issues still to be solved (Der Hessische Beauftragte für Datenschutz und Informationsfreiheit 2019b).

The first decision of the Hessian data protection authority justified the ban of Microsoft Office 365 to preserve the state's "digital sovereignty" (Der Hessische Beauftragte für Datenschutz und Informationsfreiheit 2019a, para. 2). Digital sovereignty is a concept that permeates the recent debate on data localisation in Europe widely and particularly in Germany. For example, it is the primary goal of the Gaia-X Project launched in 2019 by the German Ministry for Economic Affairs and Energy (Federal Ministry for Economic Affairs and Energy (BMWi) 2019, p. 6). In the Ministry's document, digital sovereignty is defined both as "independence" and as "self-determination" (Federal Ministry for Economic Affairs and Energy (BMWi) 2019, p. 7). Remarkably, this concept is not uniquely linked to the state dimension, encompassing also the power of companies to freely determine the use and structure of their digital systems, data and processes (Federal Ministry for Economic Affairs and Energy (BMWi) 2019, p. 7). In this way, digital sovereignty is presented as a solution to the European dependence from foreign companies and infrastructures, as well as to offer an opportunity to abide by and affirm European values.

Yet, the project of achieving European digital sovereignty is not immune from the typical criticism characterising data localisation legislation

(Mishra 2015). First, implementing a similar policy means increasing costs due to the relocation of data centres and services in Europe, and subverting global economic trends. Moreover, digital sovereignty could not be a panacea vis-à-vis the issue of security. As the Estonian project of creating a virtual data embassy shows, centralising data may enhance the level of vulnerability, while delocalisation, as the sharding procedure in the context of cloud computing services demonstrates, can actually strengthen system resilience. Lastly, initiatives aiming to preserve digital sovereignty are often criticised as ways to conceal a form of protectionism (Mishra 2015; Millard 2015; C. Kuner et al. 2015). Digital sovereignty would not merely lead to a balkanisation of the digital realm for the sake of preserving European fundamental rights, but also to allow European companies to fill the economic gap distancing them from American and Asiatic technology giants.

While Europe is seeking to strengthen its digital sovereignty, however, analogous trends are emerging also elsewhere. In 2018, for example, the US introduced the CLOUD Act, a new legislation enabling US law enforcement authorities to require US corporations to disclose data, independently of their physical location (Abraha 2019). The statute was purposefully adopted as a response to a case in which Microsoft contested a search warrant aiming to gather data stored on its Irish servers (Svantesson and Gerry 2015). Microsoft lamented that, under the Electronic Communications Privacy Act 1986, the US government was not explicitly authorised to serve extraterritorial warrants. The introduction of the CLOUD act in 2018 mooted the dispute against Microsoft, which had meanwhile reached the US Supreme Court (Abraha 2019). The new statute empowers US law enforcement authorities to require data in the 'possession, custody and control' of a US corporation, notwithstanding such information may be physically located outside the US (Abraha 2019).

Data localisation is not just a US and European phenomenon. In 2017, in the context of the increasing trade war with the US, China passed a new National Intelligence Law obliging companies to collaborate with Chinese intelligence agencies (Yang 2019). This legislation produced strong criticism in the US (Lian 2019; The White House 2019; cf. Doffman 2019). Yet it reveals a drift towards growing fragmentation of the digital space to impose national sovereignty, which raises significant challenges for cloud computing.

3.5   Conclusion

Borderless cloud computing technologies are exacerbating existing tensions between EU and US approaches to data privacy. On the one hand, a series of European initiatives are progressively exercising a centripetal force on data held by companies operating in the EU. Their main objective would be to preserve Europe's digital sovereignty by guaranteeing the respect of European fundamental rights and preventing foreign law enforcement and intelligence agencies from accessing personal data of EU citizens and residents. On the other hand, foreign countries are unilaterally adopting legislation requiring national corporations to disclose data stored in Europe, in this way bypassing jurisdictional boundaries grounded in physical data location. Both the US and Chinese recently adopted statutes represent two paradigmatic examples of this trend, and clearly highlight how a conflict between European rules and foreign laws is emerging.

From a European standpoint, it is therefore evident that data localisation alone cannot represent the universal remedy for all the existing risks. In a globalised digital environment, even investigating about a domestic crime may likely entail accessing data held in different jurisdictions. Erecting permanent barriers to the free flow of data could eventually amount to a Sisyphean labour, difficult and ultimately futile. For this reason, enhancing cooperation and establishing more functional agreements with third states, making sure that the protection of digital rights becomes a shared concern transnationally and globally, still seems to be the best choice for the EU.

References

Abraha, H. H. (2019). How Compatible Is the US "CLOUD Act" with Cloud Computing? A Brief Analysis. *International Data Privacy Law, 9*, 207–215. https://doi.org/10.1093/idpl/ipz009.

Albrecht, J. P. (2016). How the GDPR Will Change the World. *European Data Protection Law Review, 2*(3), 287–289.

Bender, D. (2016). Having Mishandled Safe Harbor, Will the CJEU Do Better with Privacy Shield? A US Perspective. *International Data Privacy Law, 6*(2), 117–138. https://doi.org/10.1093/idpl/ipw005.

Bradford, A. (2012). The Brussels Effect. *Northwestern University Law Review, 107*(1), 1–67.

Bygrave, L. A. (2014). *Data Privacy Law: An International Perspective*. Oxford: Oxford University Press.

Cate, F. H., & Cate, B. E. (2012). The Supreme Court and Information Privacy. *International Data Privacy Law, 2*(4), 255–267. https://doi.org/10.1093/idpl/ips024.

Celeste, E. (2019). The Court of Justice and the Ban on Bulk Data Retention: Expansive Potential and Future Scenarios. *European Constitutional Law Review, 15*(1), 134–157. https://doi.org/10.1017/S1574019619000038.

Cole, D., & Fabbrini, F. (2016). Bridging the Transatlantic Divide? The United States, the European Union, and the Protection of Privacy across Borders. *International Journal of Constitutional Law, 14*(1), 220–237. https://doi.org/10.1093/icon/mow012.

Council of Europe. (2019). Guide on Article 8 of the European Convention on Human Rights. Retrieved from https://www.echr.coe.int/Documents/Guide_Art_8_ENG.pdf

de Hert, P., & Czerniawski, M. (2016). Expanding the European Data Protection Scope beyond Territory: Article 3 of the General Data Protection Regulation in Its Wider Context. *International Data Privacy Law, 6*(3), 230–243. https://doi.org/10.1093/idpl/ipw008.

Der Hessische Beauftragte für Datenschutz und Informationsfreiheit. (2019a, July 9). Stellungnahme des Hessischen Beauftragten für Datenschutz und Informationsfreiheit zum Einsatz von Microsoft Office 365 in hessischen Schulen. Der Hessische Beauftragte für Datenschutz und Informationsfreiheit. Retrieved from https://datenschutz.hessen.de/service

Der Hessische Beauftragte für Datenschutz und Informationsfreiheit. (2019b, August 2). Zweite Stellungnahme zum Einsatz von Microsoft Office 365 in hessischen Schulen. Der Hessische Beauftragte für Datenschutz und Informationsfreiheit. Retrieved from https://datenschutz.hessen.de/presse-mitteilungen/zweite-stellungnahme-zum-einsatz-von-microsoft-office-365-hessischen-schulen

Digital Rights Ireland. (2014). ECLI:EU:C:2014:238. ECJ.

Doffman, Z. (2019). Trump's Huawei Ban Rejected by New Ruling in Germany. *Forbes,* 15 October. Retrieved from https://www.forbes.com/sites/zakdoffman/2019/10/15/trumps-huawei-ban-rejected-by-surprise-new-report/.

Esayas, S. Y. (2012). A Walk in to the Cloud and Cloudy It Remains: The Challenges and Prospects of "Processing" and "Transferring" Personal Data. *Computer Law & Security Review, 28*(6), 662–678. https://doi.org/10.1016/j.clsr.2012.09.007.

European Commission. (2016). *European Cloud Initiative—Building a Competitive Data and Knowledge Economy in Europe.* COM(2016) 178 Final. Retrieved from https://eur-lex.europa.eu/legal-content/EN/TXT/PDF/?uri=CELEX:52016DC0178&from=EN

European Commission. (2019). Adequacy Decisions. Text. European Commission—European Commission. Retrieved from https://ec.europa.eu/info/law/law-topic/data-protection/international-dimension-data-protection/adequacy-decisions_en

Fabbrini, F. (2015). The EU Charter of Fundamental Rights and the Rights to Data Privacy: The EU Court of Justice as a Human Rights Court. In S. de Vries, U. Bernitz, & S. Weatherill (Eds.), *The EU Charter of Fundamental Rights as a Binding Instrument* (pp. 261–286). Hart.

Federal Ministry for Economic Affairs and Energy (BMWi). (2019). Project GAIA-X—A Federated Data Infrastructure as the Cradle of a Vibrant European Ecosystem. Retrieved from https://www.bmwi.de/Redaktion/EN/Publikationen/Digitale-Welt/project-gaia-x.pdf?__blob=publicationFile&v=4

Flaherty, D. H. (1991). On the Utility of Constitutional Rights to Privacy and Data Protection. *Case Western Reserve Law Review, 41*, 831–855.

Greenleaf, G. (2019). *Global Data Privacy Laws 2019: 132 National Laws & Many Bills.* SSRN Scholarly Paper ID 3381593. Social Science Research Network, Rochester, NY. Retrieved from https://papers.ssrn.com/abstract=3381593

Greer, D. (2011). Safe Harbor—A Framework That Works. *International Data Privacy Law, 1*(3), 143–148. https://doi.org/10.1093/idpl/ipr010.

Hon, W. K., Millard, C., Singh, J., Walden, I., & Crowcroft, J. (2016). Policy, Legal and Regulatory Implications of a Europe-Only Cloud. *International Journal of Law and Information Technology, 24*(3), 251–278. https://doi.org/10.1093/ijlit/eaw006.

Hon, W. K., Hörnle, J., & Millard, C. (2012). *Data Protection Jurisdiction and Cloud Computing—When Are Cloud Users and Providers Subject to EU Data Protection Law? The Cloud of Unknowing, Part 3.* SSRN Scholarly Paper ID 1924240. Social Science Research Network, Rochester, NY. Retrieved from https://papers.ssrn.com/abstract=1924240

Hon, W. K., & Millard, C. (2012). *Data Export in Cloud Computing—How Can Personal Data Be Transferred Outside the Eea? The Cloud of Unknowing, Part 4.* SSRN Scholarly Paper ID 2034286. Social Science Research Network, Rochester, NY. Retrieved from https://papers.ssrn.com/abstract=2034286

Hon, W. K., Millard, C., & Walden, I. (2011a). *The Problem of "Personal Data" in Cloud Computing—What Information Is Regulated? The Cloud of Unknowing, Part 1.* SSRN Scholarly Paper ID 1783577. Social Science Research Network, Rochester, NY. Retrieved from https://papers.ssrn.com/abstract=1783577

Hon, W. K., Millard, C., & Walden, I. (2011b). *Who Is Responsible for "Personal Data" in Cloud Computing? The Cloud of Unknowing, Part 2.* SSRN Scholarly Paper ID 1794130. Social Science Research Network, Rochester, NY. Retrieved from https://papers.ssrn.com/abstract=1794130.

Jones, M. L. (2017). The Right to a Human in the Loop: Political Constructions of Computer Automation and Personhood. *Social Studies of Science, 47*(2), 216–239. https://doi.org/10.1177/0306312717699716.

Kuner, C., Cate, F. H., Millard, C., Svantesson, D. J. B., & Lynskey, O. (2015). Internet Balkanization Gathers Pace: Is Privacy the Real Driver? *International Data Privacy Law, 5*(1), 1–2. https://doi.org/10.1093/idpl/ipu032.

Kuner, C. (2010). Data Protection Law and International Jurisdiction on the Internet (Part 1). *International Journal of Law and Information Technology, 18*(2), 176–193. https://doi.org/10.1093/ijlit/eaq002.

Kuner, C. (2015). Extraterritoriality and Regulation of International Data Transfers in EU Data Protection Law. *International Data Privacy Law, 5*(4), 235–245. https://doi.org/10.1093/idpl/ipv019.

Lian, Y.-Z.. (2019). Opinion | Where Spying Is the Law. *The New York Times*, 13 March, sec. Opinion. Retrieved from https://www.nytimes.com/2019/03/13/opinion/china-canada-huawei-spying-espionage-5g.html

Millard, C. (2015). *Forced Localization of Cloud Services: Is Privacy the Real Driver?* SSRN Scholarly Paper ID 2605926. Social Science Research Network, Rochester, NY.

Mishra, N. (2015). *Data Localization Laws in a Digital World: Data Protection or Data Protectionism?* SSRN Scholarly Paper ID 2848022. Social Science Research Network, Rochester, NY. Retrieved from https://papers.ssrn.com/abstract=2848022

Ojanen, T. (2017). Rights-Based Review of Electronic Surveillance after Digital Rights Ireland and Schrems in the European Union. In D. Cole, F. Fabbrini, & S. Schulhofer (Eds.), *Surveillance, Privacy and Transatlantic Relations* (pp. 13–29). Hart.

Padova, Y. (2016). The Safe Harbour Is Invalid: What Tools Remain for Data Transfers and What Comes Next? *International Data Privacy Law, 6*(2), 139–161. https://doi.org/10.1093/idpl/ipw009.

Schwartz, P. M., & Solove, D. J. (2014). Reconciling Personal Information in the United States and European Union. *California Law Review, 102*, 877–916. https://doi.org/10.2139/ssrn.2271442.

Selby, J. (2017). Data Localization Laws: Trade Barriers or Legitimate Responses to Cybersecurity Risks, or Both? *International Journal of Law and Information Technology, 25*(3), 213–232. https://doi.org/10.1093/ijlit/eax010.

Solove, D. J. (2001). Digital Dossiers and the Dissipation of Fourth Amendment Privacy. *Southern California Law Review, 75*(5), 1083–1168.

Solove, D. J. (2008). *Understanding Privacy.* Cambridge, MA: Harvard University Press.

Svantesson, D., & Clarke, R. (2010). Privacy and Consumer Risks in Cloud Computing. *Computer Law & Security Review, 26*(4), 391–397. https://doi.org/10.1016/j.clsr.2010.05.005.

Svantesson, D., & Gerry, F. (2015). Access to Extraterritorial Evidence: The Microsoft Cloud Case and Beyond. *Computer Law & Security Review, 31*(4), 478–489. https://doi.org/10.1016/j.clsr.2015.05.007.

Svantesson, D. J. B. (2015). Extraterritoriality and Targeting in EU Data Privacy Law: The Weak Spot Undermining the Regulation. *International Data Privacy Law, 5*(4), 226–234. http://dx.doi.org.ucd.idm.oclc.org/10.1093/idpl/ipv024.

Tele2 Sverige. (2016). ECLI:EU:C:2016:970. ECJ.

The White House. (2019). Executive Order on Securing the Information and Communications Technology and Services Supply Chain. The White House, 15 May. Retrieved from https://www.whitehouse.gov/presidential-actions/executive-order-securing-information-communications-technology-services-supply-chain/

Tourkochoriti, I. (2014). The Snowden Revelations, the Transatlantic Trade and Investment Partnership and the Divide between U.S.-E.U. in Data Privacy Protection. *University of Arkansas at Little Rock Law Review, 36,* 161–176.

Tracol, X. (2016). EU–U.S. Privacy Shield: The Saga Continues. *Computer Law & Security Review, 32*(5), 775–777. https://doi.org/10.1016/j.clsr.2016.07.013.

Varadi, S., Kertesz, A., & Parkin, M. (2012). The Necessity of Legally Compliant Data Management in European Cloud Architectures. *Computer Law & Security Review, 28*(5), 577–586. https://doi.org/10.1016/j.clsr.2012.05.006.

Walden, I. (2011). *Accessing Data in the Cloud: The Long Arm of the Law Enforcement Agent.* SSRN Scholarly Paper ID 1781067. Social Science Research Network, Rochester, NY. Retrieved from https://papers.ssrn.com/abstract=1781067

Warren, S. D., & Brandeis, L. D. (1890). The Right to Privacy. *Harvard Law Review, 4*(5), 193–220. https://doi.org/10.2307/1321160.

Yang, Y. (2019). Is Huawei Compelled by Chinese Law to Help with Espionage? *Financial Times,* 5 March. Retrieved from https://www.ft.com/content/282f8ca0-3be6-11e9-b72b-2c7f526ca5d0

CHAPTER 4

Understanding and Enhancing Consumer Privacy Perceptions in the Cloud

Grace Fox

Abstract The recent increase in highly publicised cloud breaches, coupled with issues surrounding transparency and control in the cloud, highlights the importance of understanding and addressing privacy in this context. The extant cloud privacy literature has a tendency to focus on technical solutions to address security and privacy together, but a small emerging body of literature acknowledges the importance of consumers' privacy perceptions in the context of cloud computing. Given the breadth of cloud applications and the situational nature of privacy, it is imperative to unpack the role of privacy in this complex domain. This chapter leverages the broader privacy literature in the Information Systems field to identify potential measures to enhance consumer privacy in the cloud context and highlights a number of paths for research to further our knowledge of consumer privacy perceptions in the various cloud contexts.

G. Fox (✉)
Irish Institute of Digital Business, DCU Business School, Dublin, Ireland
e-mail: Grace.Fox@dcu.ie

© The Author(s) 2021
T. Lynn et al. (eds.), *Data Privacy and Trust in Cloud Computing*,
Palgrave Studies in Digital Business & Enabling Technologies,
https://doi.org/10.1007/978-3-030-54660-1_4

Keywords Information privacy • Data privacy • Data protection •
Cloud computing • Privacy perceptions • Cloud privacy • Cloud
security

4.1   Introduction

The cloud offers many advantages to organisations including greater efficiency and reduced data storage costs. The market for cloud computing is forecast to continue growth in 2020 with Gartner predicting a 17% increase to US$266.4 billion including an increase in the value of cloud management and security services from US$12 billion to US$13.8 billion (Gartner 2019). Today, we see cloud applications in all industries, at the consumer application level to city-wide infrastructures. However, the increasing ubiquity of cloud computing also represents new risks, not least information security and privacy vulnerabilities. Indeed, we have seen an alarming number of high-profile cloud data breaches including the largest cloud service providers. Most recently, the open Google database exposed the personal details of 200 million people (Forbes 2020). While the cloud itself is arguably more secure than physical infrastructures, human error is often the cause of these incidents. For instance, misconfiguration of cloud databases has resulted in an estimated 196 breaches from 2018–2019, leaving 33 billion records at risk, and costing organisations an estimated US$5 trillion over the two year period (DivvyCloud 2020). A single breach incident can be hugely costly to organisations. For example, Marriott could potentially be fined up to US$123 million in Europe alone for its recent cloud breach, which left the details of 5.2 million people at risk (Whittaker 2020). In addition to monetary costs, it is important to consider other implications for organisations involved in such breaches such as consumer perceptions of privacy and trust in the organisation itself.

This chapter focuses on exploring how organisations can avail of the advantages offered by the cloud, while preserving consumer privacy and addressing any privacy concerns consumers may have. The chapter proceeds with an outline of the importance of privacy and security in the cloud computing context. Next, the extant literature related to privacy in this domain and the broader Information Systems (IS) field are discussed. Potential solutions for enhancing privacy perceptions in the cloud and directions to empirically explore these solutions are outlined in the final sections of the chapter.

4.2 Cloud Computing: Privacy and Security Issues

Continual advances in information technology are furthering the proliferation of cloud computing in many new domains. The emergence of big data, and recent advances in areas such as IoT (Internet of Things) has massively increased the volumes of data generated by most organisations leading to an increasing need to outsource data storage to cloud service providers (Lowry et al. 2017). On the consumer level, the popularity and number of mobile applications downloaded by users has also resulted in a dependence on cloud computing to relieve storage issues; this is commonly referred to as mobile-cloud computing (Shropshire et al. 2015). This greater reliance on the cloud significantly exacerbates the risk of privacy and security incidents while also heightening the risks associated with more traditional security vulnerabilities (Lowry et al. 2017).

Privacy and security represent important challenges and potential barriers, both for organisations considering adopting and those currently relying on cloud services and cloud service providers (Alashoor 2014; Fauzi et al. 2012). A host of researchers have stressed the importance of addressing privacy in the cloud computing context (e.g. Pearson 2012; Wood 2012; Nikkhah et al. 2018). Indeed, the security and privacy issues within the cloud computing domain are far greater than those present when data is stored in a single location (Ramireddy et al. 2010). This is partly attributed to the fact that data stored in the cloud is often in unencrypted form and thus open to many vulnerabilities (Senarathna et al. 2016). Furthermore, the use of cloud computing often involves the movement of data beyond international borders requiring consideration of legal requirements in different jurisdictions while also complicating the organisation's ability to observe and manage data flows and preserve consumer privacy (Lowry et al. 2017). Privacy also represents an important consideration for cloud end users, with recent research illustrating that consumers are willing to pay to limit data collection and to ensure deletion from a cloud database (Shropshire et al. 2015).

While Chap. 3 in this book outlines the legal requirements across different jurisdictions, this chapter focuses on the consumer aspect of privacy in the cloud computing context. In addition to the undeniable importance of privacy in this context, it is important to note the intertwined nature of privacy and security within existing academic discussions. While, the focus here is on privacy, it is worthwhile to differentiate and highlight important parallels between these concepts. Both security and privacy have been described as human constructed abstract notions which vary according to context and other factors (Lowry et al. 2017). For the purpose of the chapter, both concepts are defined and discussed in terms of their pertinence to cloud.

4.2.1   Information Security

Information security refers to the preservation of the three tenets of security; the confidentiality of information, the integrity information, and the availability of information, while also considering other risks such as reliability, authenticity, and accountability (Pearson 2012; ISO 2005). In the cloud computing context, the key security vulnerabilities warranting consideration include trust, encryption, multi-tenancy, and reliability (Ramireddy et al. 2010). In addition, these vulnerabilities result in serious security risks related to data integrity, confidentiality, data loss, and data authentication (Subashini and Kavitha 2011). Research has provided some initial support for the relevance of these risks. For example, in their study of the factors impacting public sector cloud adoption in South Africa, Scholtz et al. (2016) found that data accessibility was a concern for 90% of participants and cyberattacks represented a concern for 76% for participants. Security and privacy are inextricably linked, as any security incident puts the privacy of the individual's data at risk (Sonehara et al. 2011). In addition, these security issues may lead to intangible risks or concerns such as loss in confidence of the reliability of the cloud and fears around access to personal data (Paquette et al. 2010).

4.2.2   Information Privacy

Privacy has been the subject of academic discourse for over two centuries in disciplines such as law, sociology and IS. Indeed, the first academic discussions of privacy are often credited to the 1890 Harvard Law Review article by Warren and Brandeis, in which they discuss privacy in terms of the need to balance individuals' rights to be free from intrusion with the information needs of society (Warren and Brandeis 1890). From a sociological perspective, the seminal definition of privacy was developed by Alan Westin (1967, p. 7), who described privacy as "the claim of individuals, groups, or institutions to determine for themselves when, how, and to what extent information about them is communicated to others." These seminal works are the building blocks of conceptualisations across multiple disciplines, many of which place control at the centre (Kesan et al. 2013). However, advances in technology have shifted the focus from a predominantly physical construct to a digital one, and from organisational control to an individual's control of their personal information. While discussions regarding the potential security and privacy issues within cloud computing are largely centered around adoption at an organisational level, there are many consumer issues which warrant consideration (Alashoor 2014). This chapter focuses on privacy from the IS perspective where privacy is defined as an individual's ability to personally control information about themselves and how it is disseminated (Smith et al. 1996; Bélanger and Crossler 2011).

In the cloud computing context, there are many privacy issues that organisations should consider and seek to address including issues surrounding control, unauthorised secondary use of data, and improper access (Senarathna et al. 2014; Pearson and Benameur 2010). However, the majority of extant privacy research in the cloud computing domain focuses on the technical solutions to secure data in the cloud from both design and architectural perspectives (Nikkhah et al. 2018). It is important to move beyond this and understand the role of the privacy perceptions of consumers on their adoption and use of cloud computing.

4.3 EXAMINING PRIVACY PERCEPTIONS IN THE CLOUD

The broader privacy literature within IS is well developed with a large number of studies conducted across multiple contexts over the past three decades. As privacy cannot be objectively measured or quantified, proxies are utilised to examine the role of privacy, with privacy concern representing the dominant approach in existing literature (Bélanger and Crossler 2011). Conceptualisations of privacy concern also differ across the myriad of existing studies with the emphasis often placed on fears around loss of privacy (Xu et al. 2011), or possible improper uses such as one's personal data being disclosed online (Son and Kim 2008). As summarized in Table 4.1 below, four scales are predominantly used to measure information privacy concerns in the IS literature (Alashoor et al. 2017). While there is an absence of agreement on the most appropriate scale, each of these scales have been rigorously tested, validated and adapted to other contexts. Furthermore, across these scales six dimensions of concern are consistently included.

Table 4.1 Dominant privacy concern scales

Authors	Scale	# dimensions	Dimensions represented					
			Collection	Unauthorised Secondary use	Improper Access	Errors	Control	Awareness
Smith et al. (1996)	Concern for Information Privacy (CFIP)	4 dimensions	✓	✓	✓	✓	x	x
Malhotra et al. (2004)	Internet Users' Information Privacy Concerns (IUIPC)	3 dimensions	✓	x	x	x	✓	✓
Dinev and Hart (2006)	Privacy Concerns (PC)[a]	Unidimensional	✓	✓	✓	x	x	x
Hong and Thong (2013)	Internet Privacy Concerns (IPC)	6 dimensions	✓	✓	✓	✓	✓	✓

[a]This measure is unidimensional in nature with items representing three dimensions as noted above.

While these privacy measures are commonly deployed in other contexts, the majority of extant cloud computing privacy studies do not use validated measures of privacy concern but instead explore privacy and security issues together using open-ended questions or single-item ranking questions. For example, in a study conducted by Scholtz et al. (2016), 90% of participants rated the privacy of data as important or very important. However, two related studies adapted Dinev and Hart's (2006) PC measure (Nikkhah et al. 2018; Nikkhah and Sabherwal 2017). Validated measures of information privacy concern warrant consideration in future cloud computing privacy studies to provide a more nuanced view of privacy in the cloud and to allow comparisons to be drawn with privacy concerns in other contexts. Indeed, many of these dimensions represent core privacy issues highlighted by cloud researchers (e.g. Pearson and Benameur 2010). However, the extant empirical literature has not yet encompassed these dimensions. The relevance of these dimensions is briefly noted in terms of understanding consumers' perceptions of privacy.

The collection dimension focuses on individuals' concerns regarding an organisation's collection and storage of a great deal of their personal information (Smith et al. 1996). Consumers often lack an awareness of how their data stored in the cloud is used and disseminated, and whether it is used for purposes other than those it was collected for (Nikkhah et al. 2018). For example, in some cases such as Google Drive or Dropbox, the storage of personal information in the cloud is the primary purpose of the service and therefore use is transparent. Other applications such as those in the Internet of Things (IoT) domain, are less clear. Data may be stored on the device, somewhere locally, or in the cloud, or a combination of one or more these. Consumers may not even be aware of where data is stored. It is important to explore whether cloud data storage generates consumer privacy concern and how this differs across applications and information types.

The Unauthorised Secondary Use dimension focuses on individuals' concerns that information collected for one purpose is subsequently used for a secondary purpose without obtaining the individual's permission (Smith et al. 1996). Consumer perceptions of unauthorised secondary use in the cloud context are highlighted in extant research (Pearson and Benameur 2010). The Improper Access dimension covers individuals' concerns that an organisation does not have the measures in place to prevent unauthorised individuals from accessing their information (Smith et al. 1996). The recent media coverage around large cloud data breaches may heighten consumer awareness of potential improper access to their data stored in the cloud, and consequently, increase their concerns around such access. The Errors measure of concern focuses on individuals' concerns that the organisation storing their personal information does not have the measures in place to prevent and correct errors in the data (Smith et al. 1996). This dimension may not be relevant in all consumer cloud contexts, but issues around controlling data flows which are inherent in the cloud may cause concern around organisations' ability to track information, and as a result their capabilities to identify and remedy errors.

The Control dimension focuses on individuals' concerns regarding the lack of control they have over their personal information (Malhotra et al. 2004). Issues around control over data has been highlighted as an important cloud privacy issue that warrants exploration (Sun et al. 2011; Pearson and Benameur 2010). The Awareness dimension centres around individuals' concerns regarding their lack of awareness of how an organisation uses and protects their personal information (Malhotra et al. 2004). Awareness represents another core concern in the cloud context, with lack of transparency around where data is stored and the protection mechanisms in place (e.g. Singh et al. 2015). It is important to therefore examine (1) if consumers lack awareness of privacy practices in the cloud and (2) if a lack of awareness heightens concerns for the privacy of one's personal information.

The broader privacy literature offers a rich theoretical base from which the role of privacy in cloud computing can be advanced. Privacy theories are typically discussed from five perspectives—(1) drivers of privacy concern, (2) behavioural consequences, (3) trade-offs, (4) institutional drivers and (5) individual factors (Li 2012). In contrast to the broader literature, the privacy research in the cloud context is relatively nascent. A review of the literature did not identify any studies leveraging theories in four of these five categories. However, two related studies drew on privacy calculus theory (PCT), a theory commonly utilised to understand the trade-offs between the benefits and the risks associated with the behaviour in question such as information disclosure or using a new technology (Culnan and Armstrong 1999). PCT posits that individuals engage in a cognitive comparison of the benefits and potential negative outcomes which may result from using a certain technology (Culnan and Armstrong 1999). According to this theory, individuals will utilise the technology as long as their perceptions of the benefits outweigh their risk perceptions (Culnan 1993). The first study focused on consumers' willingness to disclose information in cloud-based mobile applications and found that privacy concerns reduced willingness to disclose information, whereas perceived usefulness of the apps and perceived trustworthiness both positively influenced willingness to disclose (Nikkhah and Sabherwal 2017), thereby supporting the use of PCT in this context. The second study focused on consumers' intentions to continue use of cloud-based mobile applications and found that security and privacy interventions reduce privacy concerns and increase trust, and privacy and security concerns both indirectly influence willingness to continue to use apps through trust (Nikkhah et al. 2018), furthering support for PCT. These studies support the potential of privacy theories to advance our understanding of the role of privacy in this context and point to the need to further leverage this rich theoretical base.

4.4   Enhancing Privacy Perceptions in the Cloud

The focus of much of the privacy literature in the cloud domain has been on technical measures to secure data to enhance security in the hope of negating privacy issues. These are discussed further in Chap. 7 of this book. These studies focus on reviewing the efficacy of measures such as different approaches and anonymisation mechanisms (Sonehara et al. 2011). Recent works have also highlighted important security considerations in emerging cloud contexts such as IoT, stressing the need to consider secure communications, data identification measures, and certification approaches for example (Singh et al. 2015).

Drawing from the existing research in the cloud context and the broader privacy literature, it can be argued that organisations need to address three related consumer perceptions; (1) control (2) awareness and (3) trust. The first two perceptions relate specifically to privacy concern while the third represents a broader perception of the technology (as discussed in detail in Chap. 1). All three consumer perceptions can negatively influence individuals' willingness to adopt new technologies (Li 2012) and as such, represent an important barrier to the continued success of cloud computing.

Perceived control is a primary concept within the information privacy literature. However, it is important to note that privacy and control are conceptually distinct (Laufer and Wolfe 1977) and negatively correlated. Control is a perception based variable and has been defined as an individual's beliefs in their ability to manage the collection and use of their personal data (Xu et al. 2011). In the cloud context, it has been noted that consumers are afforded little or no control over their information (Alashoor 2014). In other contexts, a lack of perceived control can heighten privacy concerns (Dinev and Hart 2004), whereas if perceived control is high, individuals may be empowered to adopt technologies and disclose more personal information (Palmatier and Martin 2019). Closely related to control is consumers' awareness of how their information is protected and used in the cloud. A lack of transparency is a commonly cited issue in the cloud context with many noting that cloud providers should engage in transparent communications to increase consumer awareness of how their personal data is used (Kesan et al. 2013). Awareness not only encompasses understanding of how data is protected, but where data is stored and how it is used.

Trust is often incorporated into privacy studies including those in the cloud context (e.g. Nikkhah et al. 2018). Consumers' beliefs regarding the trustworthiness of an organisation relate to perceptions of the organisation's benevolence, integrity and competence (van der Werff et al. 2019; Bélanger et al. 2002). In privacy contexts, trust often focuses on an individual's willingness to be vulnerable when transacting or sharing personal information with an organisation (McKnight et al. 2002). In the cloud context, research supports the importance of trust in influencing consumers' willingness to use cloud-based mobile applications and the relationship between privacy concern and trust (Nikkhah et al. 2018).

Central to addressing these consumer perceptions is improving organisational communications with consumers and building knowledge. Organisation's current communication efforts largely involve privacy policies. Currently, privacy policies tend to be quite lengthy and difficult to read (Kelley et al. 2010). Indeed, the time to read the privacy policies of all websites visited by an average American Internet user was estimated as 201 hours annually (McDonald and Cranor 2008). Furthermore, when consumers read privacy policies, they often do not understand the contents (Martin 2015). These issues with readability and understandability, as well as lack of user engagement with privacy policies, has led to calls to develop new communication methods which better inform consumers how their information is used (Park et al. 2012). In addition to how organisations communicate, it is important to ensure consumers are equipped with the privacy knowledge needed to interpret these communications. Indeed, gaps in consumers' privacy knowledge and self-efficacy has recently been highlighted as an important area to address in order to empower informed decision making (Crossler and Bélanger 2017). Thus, we present three approaches organisations can use to influence consumer perceptions related to privacy in the cloud namely institutional assurances, just-in-time interventions, and building privacy knowledge. The first two approaches directly relate to communication methods and the third approach focuses on building consumers' privacy literacy and as a result their capacity to engage with organisation consumers regarding their privacy practices.

4.4.1 Institutional Assurances

Institutional assurances or privacy disclosures are communication efforts from organisations to consumers, regarding the organisation's data privacy practices. Institutional assurances are often heralded as a solution to addressing privacy concerns, improving perceptions of control and enhancing trust beliefs in many contexts (Culnan and Armstrong 1999; Wu et al. 2012). Institutional assurances include privacy policies and visual approaches which combine text and icons such as privacy labels and trust labels. The privacy literature has found effective privacy policies address awareness issues by increasing perceptions of control (Xu et al. 2011) and improving understanding of privacy practices (Kelley et al. 2010). However, the weaknesses inherent in existing privacy policies led to the emergence of the nutritional privacy label (Kelley et al. 2010). In the cloud context, a cloud trust label providing relevant institutional assurances and privacy information has been found to impact decision makers' perceptions of the trustworthiness of cloud service providers (van der Werff et al. 2019). The nutritional label is one possible approach to institutional assurances which could serve as a fruitful avenue for both cloud service providers and organisations leveraging the cloud. This approach should include all information required in GDPR compliant privacy notices and inform consumers of how their data is used, stored, protected, and the controls they can exercise over their data. In addition, the label should address the three dimensions of trust (benevolence, integrity and competence), and the core security considerations in the cloud context. The label approach combines many recommendations for effective communication from the UK Information Commissioner's Office (ICO) including the use of recognisable icons and layering formats (ICO 2017). The content of the label should differ depending on the organisation and whether the label is consumer facing or used to influence perceptions of key decision makers in organisations. Based on findings in the cloud and other contexts, these labels may build awareness and strengthen perceptions of control and trustworthiness. Privacy labels are likely to be useful as a communication tool for prospective customers, and should be accessible to existing customers within the application or website settings.

4.4.2   Just-in-Time Interventions

In addition to detailed communication approaches such as privacy labels and notices, there is also a need for additional transparent communications with consumers as the need arises. The ICO advocates the use of just-in-time notices to inform consumers of changes in an organisation's privacy practices (ICO 2017). For example, if an organisation plans to migrate to the cloud, they should inform their consumers of this change in a transparent manner. Additional reasons for just-in-time interventions include cases when an additional use arises for personal data and consumer consent is sought, as well as requests for updates to personal information, or requests for more information. In these interventions, organisations should be transparent and focus on explaining why the data is needed. The format of these interventions will vary depending on the level of the change and the technology in question. For mobile applications, a pop-up notification could be utilised to request consumer consent or additional data disclosure. For websites, individuals could be prompted to provide the requested information at log-in. The purpose of just-in-time interventions is similar to the privacy label approach in that they seek to overcome issues with awareness, while reminding consumers of the controls they have and seeking to enhance trust beliefs through transparency.

4.4.3   Building Privacy Knowledge

As repeatedly noted throughout this chapter, cloud consumers often lack awareness of how their data is stored and used. In the broader privacy literature, Crossler and Bélanger (2019) advocate the importance of understanding and addressing consumers' privacy knowledge-belief gaps and the need to develop contextualised privacy self-efficacy i.e. individuals' perceptions that they have the knowledge and skills needed to protect the privacy of their data as required. This work builds on findings around the privacy paradox, wherein consumers express high privacy concerns but do not engage in behaviours to protect their information privacy (Bélanger and Crossler 2011). In their recent study, Crossler and Bélanger (2019) found that context-specific privacy knowledge and privacy self-efficacy influence individuals' privacy-protective behaviours. There is a need for organisations to consider consumers' privacy knowledge and self-efficacy both for potential and existing customers. Any communication, be it a privacy label or app notification regarding privacy should consider the consumers' knowledge, and be framed in ways which enhance their knowledge as opposed to obfuscate improper practices. In addition, organisations should provide supplementary information and resources which empower consumers to develop their privacy knowledge and self-efficacy. It is argued that this level of transparency will heighten perceptions of control and trustworthiness while building consumers' privacy self-efficacy and thereby facilitating informed decision making.

4.5   Future Research Directions

As discussed earlier, consumer perceptions of privacy are influenced by past experience and the context in question (Li 2011). As such, pertinent privacy issues are likely to vary across different cloud contexts (Pearson 2012). At a high level, more privacy research is required that focuses on consumer perceptions of privacy in the cloud in general, among different cloud service provision models, in public/private cloud settings, and in different domains such as IoT. Table 4.2 below outlines a number of potentially fruitful directions for research that may enrich our understanding of privacy in the cloud. As per Li (2012), relevant theories popularised in other privacy contexts are listed alongside each area. These research directions represent an initial step in unravelling the role of the complex privacy construct in this multifaceted and evolving context.

4.6   Concluding Remarks

The need to address privacy concerns in order to ensure the success of new information technologies has been argued in the broader privacy literature (Hong and Thong 2013). The importance of addressing privacy and security in the cloud is also well established (Wood 2012). However, literature focused on understanding consumer perceptions regarding privacy in the cloud is still emerging. This chapter argues for the need to move beyond technical solutions which address security first and privacy second, towards a focus on understanding and addressing the privacy perceptions of consumers. Given the proliferation of cloud computing, the potential privacy

Table 4.2 Future research directions

Focus	Areas of focus	Guiding theories	Rationale
Consumer privacy perceptions	• Examining the Antecedents to privacy concern across differing cloud applications. • Unravelling the outcomes of privacy concern in different contexts including adoption, continuance and disclosure behaviours. • Further unravelling the trade-offs between perceived benefits and risks in different cloud domains. • Examining the role of important privacy-related constructs including trust and risk.	• Privacy calculus theory • Social contract theory • Protection Motivation theory	Much of the privacy literature in other contexts follows the APCO model developed by Smith et al. (2011) to explore the antecedents, privacy concerns and outcomes in a given context. Little is known about the important drivers of concern, dimensions of concern or outcomes in the cloud context.
Effectiveness of institutional assurances	• Developing privacy labels and just-in-time interventions for different cloud applications and customers e.g. end users and business decision makers. • Examining the influence of these assurances on potential consumers' perceptions of privacy and trust. • Examining the influence of these assurances on existing customers perceptions and willingness to continue use.	• Communication privacy management theory • Procedural fairness theory	The potential of institutional assurances to influence consumer perceptions and behaviours is documented in other contexts. There is a need to unravel their effectiveness in the cloud context given the breadth of applications and technologies relying on the cloud to some degree to deliver services.

(*continued*)

Table 4.2 (continued)

Focus	Areas of focus	Guiding theories	Rationale
Role of regulation	• Examining consumer perceptions of regulatory protections in the cloud context. • Investigating regulatory expectations across cultures and cloud contexts.	• Regulatory focus theory	The issue of data location is a primary concern in the cloud context as this has implications for regulation. In addition, perceptions regarding regulation has been found to influence privacy in other contexts (Li 2011). Little is known in terms of consumers' regulation knowledge and expectations in the cloud context.
Privacy and trust repair	• Examine the impacts of cloud privacy breaches on consumer perceptions regarding the cloud generally and the affected organisation. • Explore the efficacy of breach responses in rebuilding trust and privacy perceptions	• Psychological contract breach • Communication privacy management theory	The growth of privacy breaches in the cloud has been widely publicised. It is important to explore how these incidents influence consumer perceptions and how organisations can recover.
Building privacy knowledge	• Explore consumers existing privacy knowledge and self-efficacy in different cloud contexts. • Examine the efficacy of different knowledge and self-efficacy building approaches. • Explore the relationship between self-efficacy, privacy concern and behaviours in different cloud contexts.	• Self-efficacy theories • Information-Motivation-Belief theory	Consumers lack of awareness of how their data is used in the cloud context coupled with the growth of cloud data breaches point to the need to understand and develop privacy knowledge and self-efficacy among consumers. As a result, it is important to empirically explore the efficacy of any such knowledge building efforts.

implications span multiple industries, and privacy may pose different challenges for each industry or application. Proactive approaches to communicate with consumers such as privacy labels can be useful in addressing privacy concerns, enhancing perceptions of control and building trust beliefs. In addition, efforts are needed to build the privacy literacy and self-efficacy of consumers in this context. The recommendations presented in this chapter emphasises the importance of organisations proactively understanding and positively influencing consumer privacy perceptions, over and above the compliance with legal requirements such as the GDPR. It is hoped that this chapter provides some useful recommendations for practice and presents some interesting avenues for research in this domain.

REFERENCES

Alashoor, T. (2014). Cloud Computing: A Review of Security Issues and Solutions. *International Journal of Cloud Computing, 3*(3), 228–244.

Alashoor, T., Fox, G., & Jeff Smith, H. (2017). *The Priming Effect of Prominent IS Privacy Concerns Scales on Disclosure Outcomes: An Empirical Examination.* Pre-ICIS Workshop on Information Security and Privacy.

Bélanger, F., & Crossler, R. E. (2011). Privacy in the Digital Age: A Review of Information Privacy Research in Information Systems. *MIS Quarterly, 35*(4), 1017–1042.

Bélanger, F., Hiller, J. S., & Smith, W. J. (2002). Trustworthiness in Electronic Commerce: The Role of Privacy, Security, and Site Attributes. *The Journal of Strategic Information Systems, 11*(3–4), 245–270.

Crossler, R. E., & Bélanger, F. (2017). *The Mobile Privacy-Security Knowledge Gap Model: Understanding Behaviors.* Proceedings of the 50th Hawaii International Conference on System Sciences.

Crossler, R. E., & Bélanger, F. (2019). Why Would I Use Location-Protective Settings on My Smartphone? Motivating Protective Behaviors and the Existence of the Privacy Knowledge–Belief Gap. *Information Systems Research, 30*(3), 995–1006.

Culnan, M. J. (1993). "How Did They Get My Name?" An Exploratory Investigation of Consumer Attitudes toward Secondary Information Use. *MIS Quarterly, 17*, 341–363.

Culnan, M. J., & Armstrong, P. K. (1999). Information Privacy Concerns, Procedural Fairness, and Impersonal Trust: An Empirical Investigation. *Organization Science, 10*(1), 104–115.

Dinev, T., & Hart, P. (2004). Internet Privacy Concerns and Their Antecedents-Measurement Validity and a Regression Model. *Behaviour & Information Technology, 23*(6), 413–422.

Dinev, T., & Hart, P. (2006). An Extended Privacy Calculus Model for E-Commerce Transactions. *Information Systems Research, 17*(1), 61–80.

DivvyCloud. (2020). 2020 Cloud Misconfigurations Report. Retrieved from https://divvycloud.com/misconfigurations-report-2020/

Forbes. (2020). Beware—This Open Database On Google Cloud 'Exposes 200 Million Americans': Are You at Risk?. Retrieved from https://www.forbes.com/sites/zakdoffman/2020/03/20/stunning-new-google-cloud-breach-hits-200-million-us-citizens-check-here-if-youre-now-at-risk/#1737e9528587

Fauzi, A. A. C., Noraziah, A., Herawan, T., & Zin, N. M. (2012, March). On cloud computing security issues. In *Asian Conference on Intelligent Information and Database Systems* (pp. 560–569). Springer, Berlin: Heidelberg.

Gartner. (2019). Gartner Forecasts Worldwide Public Cloud Revenue to Grow 17% in 2020. Retrieved from https://www.gartner.com/en/newsroom/press-releases/2019-11-13-gartner-forecasts-worldwide-public-cloud-revenue-to-grow-17-percent-in-2020

Hong, W., & Thong, J. Y. L. (2013). Internet Privacy Concerns: An Integrated Conceptualization and Four Empirical Studies. *MIS Quarterly, 37*(1), 275–298.

ICO. (2017). Privacy Notices, Transparency and Control. A Code of Practice on Communicating Privacy Information to Individuals. Retrieved from https://ico.org.uk/for-organizations/guide-to-dataprotection/privacy-notices-transparency-and-control/

ISO 27001. (2005). Information Security Management—Specification with Guidance for Use.

Kelley, P. G., Cesca, L., Bresee, J., & Cranor, L. F. (2010). *Standardizing Privacy Notices: An Online Study of the Nutrition Label Approach.* Proceedings of the SIGCHI Conference on Human Factors in Computing Systems, pp. 1573–1582.

Kesan, J. P., Hayes, C. M., & Bashir, M. N. (2013). Information Privacy and Data Control in Cloud Computing: Consumers, Privacy Preferences, and Market Efficiency. *Washington and Lee Law Review, 70*, 341.

Laufer, R. S., & Wolfe, M. (1977). Privacy as a Concept and a Social Issue: A Multidimensional Developmental Theory. *Journal of Social Issues, 33*(3), 22–42.

Li, Y. (2011). Empirical Studies on Online Information Privacy Concerns: Literature Review and an Integrative Framework. *CAIS, 28*, 28.

Li, Y. (2012). Theories in Online Information Privacy Research: A Critical Review and an Integrated Framework. *Decision Support Systems, 54*(1), 471–481.

Lowry, P. B., Dinev, T., & Willison, R. (2017). Why Security and Privacy Research Lies at the Centre of the Information Systems (IS) Artefact: Proposing a Bold Research Agenda. *European Journal of Information Systems, 26*(6), 546–563.

Malhotra, N. K., Kim, S. S., & Agarwal, J. (2004). Internet Users' Information Privacy Concerns (IUIPC): The Construct, the Scale and a Causal Model. *Information Systems Research, 15*(4), 336–355.

Martin, K. (2015). Privacy Notices as Tabula Rasa: An Empirical Investigation Into How Complying with a Privacy Notice is Related to Meeting Privacy Expectations Online. *Journal of Public Policy & Marketing, 34*(2), 210–227.

McDonald, A. M., & Cranor, L. F. (2008). The Cost of Reading Privacy Policies. *Information System: A Journal of Law and Policy for the Information Society, 4*(3), 543–568.

McKnight, D. H., Choudhury, V., & Kacmar, C. (2002). Developing and Validating Trust Measures for e-commerce: An Integrative Typology. *Information Systems Research, 13*(3), 334–359.

Nikkhah, H. R., & Sabherwal, R. (2017). *Mobile Cloud-Computing Applications: A Privacy Cost-Benefit Model.* Proceedings of the Twenty-third Americas Conference on Information Systems, Boston.

Nikkhah, H. R., Grover, V., & Sabherwal, R. (2018). *Why Do Users Continue to Use Mobile Cloud Computing Applications? A Security-Privacy.* Proceedings of the 13th Pre-ICIS Workshop on Information Security and Privacy, Vol. 1.

Palmatier, R. W., & Martin, K. D. (2019). *The Intelligent Marketer's Guide to Data Privacy.* Springer.

Paquette, S., Jaeger, P. T., & Wilson, S. C. (2010). Identifying the Security Risks Associated with Governmental Use of Cloud Computing. *Government Information Quarterly, 27*(3), 245–253.

Park, Y. J., Campbell, S. W., & Kwak, N. (2012). Affect, Cognition and Reward: Predictors of Privacy Protection Online. *Computers in Human Behavior, 28*(3), 1019–1027.

Pearson, S. (2012). *Privacy, Security and Trust in Cloud Computing,* pp. 1–57. Retrieved from https://www.hpl.hp.com/techreports/2012/HPL-2012-80R1.pdf

Pearson, S., & Benameur, A. (2010, November). *Privacy, Security and Trust Issues Arising from Cloud Computing.* 2010 IEEE Second International Conference on Cloud Computing Technology and Science (pp. 693–702). IEEE.

Ramireddy, S., Chakraborty, R., Raghu, T. S., & Rao, H. R. (2010). Privacy and Security Practices in the Arena of Cloud Computing-A Research in Progress. In *AMCIS,* p. 574.

Scholtz, B., Govender, J., & Gomez, J. M. (2016). Technical and Environmental Factors Affecting Cloud Computing Adoption in the South African Public Sector. *CONF-IRM,* p. 16.

Senarathna, I., Yeoh, W., & Warren, M. (2014). Security and Privacy Concerns for Australian SMEs Cloud Adoption. *WISP 2014 Proceedings.*

Senarathna, I., Yeoh, W., Warren, M., & Salzman, S. (2016). Security and Privacy Concerns for Australian SMEs Cloud Adoption: Empirical Study of

Metropolitan vs Regional SMEs. *Australasian Journal of Information Systems, 20*, 1–20.

Shropshire, J., Campbell, M., & Bob, S. (2015). *Overcoming Privacy Challenges in Mobile-Cloud Computing.* SAIS 2015 Proceedings, p. 28

Singh, J., et al. (2015). Twenty Security Considerations for Cloud-Supported Internet of Things. *IEEE Internet of things Journal, 3*(3), 269–284.

Smith, H. J., Dinev, T., & Xu, H. (2011). Information Privacy Research: An Interdisciplinary Review. *MIS Quarterly, 35*(4), 989–1016.

Smith, H. J., Milberg, S. J., & Burke, S. J. (1996). Information Privacy: Measuring Individuals' Concerns about Organizational Practices. *MIS Quarterly, 20*, 167–196.

Son, J.-Y., & Kim, S. S. (2008). Internet Users' Information Privacy-Protective Responses: A Taxonomy and a Nomological Model. *MIS Quarterly, 32*(3), 503–529.

Sonehara, N., Echizen, I., & Wohlgemuth, S. (2011). Isolation in Cloud Computing and Privacy-Enhancing Technologies. *Business & Information Systems Engineering, 3*(3), 155.

Subashini, S., & Kavitha, V. (2011). A Survey on Security Issues in Service Delivery Models of Cloud Computing. *Journal of Network and Computer Applications, 34*(1), 1–11.

Sun, D., Chang, G., Sun, L., & Wang, X. (2011). Surveying and Analysing Security, Privacy and Trust Issues in Cloud Computing Environments. *Procedia Engineering, 15*, 2852–2856.

van der Werff, L., Fox, G., Masevic, I., Emeakaroha, V. C., Morrison, J. P., & Lynn, T. (2019). Building Consumer Trust in the Cloud: An Experimental Analysis of the Cloud Trust Label Approach. *Journal of Cloud Computing, 8*(1), 6.

Warren, S. D., & Brandeis, L. D. (1890). The Right to Privacy. *Harvard Law Review*, 193–220.

Westin, A. (1967). *Privacy and Freedom.* New York: Atheneum.

Whittaker, Z. (2020). *Marriott Says 5.2 Million Guest Records Were Stolen in Another Data Breach.* TechCrunch. Retrieved from https://techcrunch.com/2020/03/31/marriott-hotels-breached-again/

Wood, K. (2012). Exploring Security Issues in Cloud Computing. *UKAIS.*

Wu, K.-W., Huang, S. Y., Yen, D. C., & Popova, I. (2012). The Effect of Online Privacy Policy on Consumer Privacy Concern and Trust. *Computers in Human Behavior, 28*(3), 889–897.

Xu, H., et al. (2011). Information Privacy Concerns: Linking Individual Perceptions with Institutional Privacy Assurances. *Journal of the Association for Information Systems, 12*(12), 798–824.

CHAPTER 5

Justice vs Control in Cloud Computing: A Conceptual Framework for Positioning a Cloud Service Provider's Privacy Orientation

Valerie Lyons

Abstract The continued rise in frequency and magnitude of cloud-based privacy breaches brings to the fore the challenges experienced by cloud service providers (CSPs) in balancing the need to maximize profit with the need to maintain data privacy. With a backdrop of the ineffectiveness of regulatory approaches to protecting privacy, this chapter explores privacy from a non-regulatory perspective—instead exploring a CSP's approach to privacy as dynamics of control and justice. We apply control theory to represent the CSP's compliance with privacy legislation and power over data, and we apply justice theory to represent the CSP exceeding compliance. Control theories, such as social contract theory, have frequently

V. Lyons (✉)
The Linc Centre, BH Consulting, Blanchardstown Technology University, Dublin, Ireland
e-mail: valerie.lyons3@mail.dcu.ie

T. Lynn et al. (eds.), *Data Privacy and Trust in Cloud Computing*,
Palgrave Studies in Digital Business & Enabling Technologies,
https://doi.org/10.1007/978-3-030-54660-1_5

been applied to explore privacy challenges between organizations and consumers, as too have justice theories e.g. procedural and distributive justice. However, few studies have combined these theoretical concepts to provide a balanced view of these tensions in the cloud computing landscape. Integrating concepts from these theories, we construct a framework that can help to explain and position a CSP's privacy orientation. Four key privacy orientations emerge in our framework, namely: Risk Managers, Integrators, Citizens and Warriors. We discuss the implications of each privacy orientation for CSPs. Our framework will enable future research to further understand, explore and compare the impact and effectiveness of each privacy orientation.

Keywords Cloud computing • Data privacy • Data protection • Control theories • Justice theories • Procedural justice • Distributive justice

5.1   Introduction

The high cost of maintaining internal IT systems that are scalable, robust, and fast enough to keep pace with the speed of business, has ushered in the era of cloud computing services. Global cloud market revenues are predicted to increase from U$180b in 2015 to U$390b in 2020, attaining 17% annual average growth (Forbes 2017). All indicators for the growth of cloud computing, such as data traffic, the number of data centers or the amount of cloud service providers (CSPs) are also predicted to grow exponentially (Cisco 2018a). Whether applied to enterprise management systems, email systems or mobile computing, cloud computing both increases efficiency and reduces information technology costs by allowing on-demand access to a shared pool of computing resources that can be rapidly provisioned and released with minimal management effort (Mell and Grance 2011). However, although cloud computing's benefits are tremendous, security and privacy concerns continue to be the primary obstacles to wide adoption (CSA 2009).

Cloud computing presents particularly challenging issues for privacy as it involves the dispersal of data across servers in geographically dispersed locations that often cross national and legislative boundaries, together with shared use of resources, making privacy incidents difficult to determine and detect (Ren et al. 2012). This was evidenced by a recent case in 2019, when Rubrik, the cloud data management giant, exposed a large cache of customer information improperly stored in an Amazon Elasticsearch database (Techcrunch 2019). With the rise in high-profile privacy incidents, such as Cambridge Analytica in 2018, and Capital One's data breach on AWS (Fortune 2019), cloud service customers (CSCs) have also become more aware of the potential risks associated with the processing of data outside the traditional on-premise models. There are additional tensions arising from the costs of detection, prevention and/or remediation of privacy incidents versus the value that can be derived from data services (Acquisti 2008; Chen et al. 2012).

Governments, regulators and policymakers respond to such incidents by enhancing the specificity and stringency of compliance regulation, however the pace of change in cloud computing and the exponential growth in data is fast outpacing the legislative lifecycle. Instead of focusing on regulation as a response, focusing on privacy from a non-regulatory perspective may yield more sustainable solutions for CSPs to address the balance of these tensions more effectively. Reflecting this—in this chapter, we draw on control theories of privacy (Fried 1984; Moor 1997) to explain how CSPs comply with privacy law and exact power over data and systems, and we draw on theories of procedural justice (Allan and Tyler 1988) to explain how CSPs can 'loosen the leash' to enable the CSC have more control.

It would be highly complex to map cloud computing issues across the full panoply of regulatory privacy architectures, such as the California Consumer Privacy Act (CCPA), the General Data Protection Regulation (GDPR), the United Nations Declaration of Human Rights (UDHR), the Health Insurance Portability and Accountability Act (HIPAA) etc. However, given the broad reach of GDPR (it jurisdictionally applies to all EU CSPs, and to any non-EU CSP processing data of EU residents, or to any data processing within an EU territory) and its position as the strongest data protection regime in the world (Consumers-International 2019), we focus on GDPR in this chapter as the common reference for privacy regulation.

Applying these concepts of control and justice to privacy, we present a proposed privacy orientation framework describing the key privacy orientations of CSPs. This framework can help explain different approaches to privacy, and to understand their implications. Our framework extends research from Greenaway et al. (2015) to posit a new dimension to privacy orientations, which we call the Philanthropic Privacy dimension, where organizations undertake privacy activities with the aim of resolving privacy as a social issue. This privacy dimension is important, as organizations taking actions beyond their corporate obligations have been found to experience less privacy incidents (Accenture and Ponemon 2015).

The remainder of this chapter is organized as follows: in the next section we discuss the tensions arising from privacy, through the lens of control and justice. Following that section, we provide a brief overview of privacy risks in the CSP environment and outline the boundaries of privacy responsibilities within cloud computing models. We subsequently describe the development of our privacy orientation framework. We conclude this chapter with proposals for application of the framework and suggestions for further research.

5.2   Privacy as a Control or Justice Behavior

Dinev (2014) suggests that societal discourse has come to equate privacy and information privacy, so the terms are used here interchangeably. Information privacy is frequently described as a multidimensional concept that is dependent on context (Culnan and Williams 2009; Xu et al. 2012). The multiple dimensions affecting privacy being harm, protection, provision, and scope (Mulligan et al. 2016) and the facets shaping context being information sensitivity, industry sector, political sector and technological applications (Smith et al. 2011).

For over two decades, theories of control or theories of justice have been used to explore privacy challenges. Control theories, such as social contract theory, have frequently been applied to explore privacy challenges between organizations and consumers (Martin 2012, 2016; Wright and Xie 2019) as too have justice theories e.g. procedural and distributive justice (Ashworth and Free 2006; Culnan and Armstrong 1999). However, with the exception of Greenaway et al. (2015) no other study has combined these theoretical concepts to provide a balanced view of the privacy tensions arising between CSPs and CSCs.

5.2.1   Privacy as Control

While there is no single concept of information privacy that crosses all disciplines (Smith et al. 2011) 'control over personal information' is a common theme across many privacy studies (Belanger and Crossler 2011; Belanger et al. 2002). Control can refer to the 'controls' used to manage privacy (Belanger et al. 2002) or refer to the dynamic of 'power' over data (Johnson 2009). Both CSCs and CSPs implement 'controls' to manage privacy (Belanger et al. 2002) in the form of privacy enhancing tools, network access controls, authorization and authentication controls, privileged identity management controls etc. When privacy controls fail, a privacy incident is said to occur (a privacy incident is defined as the loss of control, compromise, unauthorized disclosure, unauthorized acquisition, or any similar occurrence where an authorized user accesses or potentially accesses personal information (DHS 2017, p. 8)). Johnson (2009) on the other hand refers to control as an organization's need for power over information, while the consumer has to balance the good achieved through information processing against their need for privacy.

Providing the CSC with increased control over their data can result in the CSP not being able to maximize data storage efficiency, require the implementation of costly technical tools by the CSP, or necessitate costly legal indemnification of risk away from the CSP. Greenaway et al. (2015) describe this tension as the pursuit of interests such as profitability or market share, at the expense of those who provide information or pay for a service. This need for a CSP to dominate control is enshrined in the concept of information ownership, and it is this concept of control and power that we apply in this chapter, as reflected by Xu et al. (2012) who distinguish privacy control between individual and organization, with the organization increasingly becoming the 'control agent'.

However, privacy is not solely about control but also about information being authorized to flow to specific agents at specific times (Moor 1997). Moor (1997) argues that in a highly digital culture it is simply not possible to control all information. Therefore, he argues, the best way to protect privacy is to ensure the right people have access to relevant information at the right time, giving individuals as much control over their data as realistically possible (labelling his theory the "control/restricted access" theory of privacy). With the emerging complexity of organizational networks, such as those constructed by CSPs, such levels of control are not realistically possible. Greenaway et al. (2015) classify organizations who provide little control to their consumers as 'low control'. However, aligning to the concepts of control that we draw on in this paper, we would classify these organizations as 'high control', as they essentially dominate 'power' over the consumers' information.

5.2.2   Privacy as Justice

The power-responsibility equilibrium (PRE) model, developed by Davis et al. (1980), states that power (which Laczniak and Murphy (1993) define as 'the ability to control') should be in equilibrium, where the partner with more power also has the responsibility to ensure an environment of trust and confidence. If an organization chooses a strategy of great power and less responsibility, it might benefit in the short-term but will lose control in the long-term e.g. from increased regulation (Caudill and Murphy 2000). Organizations can re-balance the control-equilibrium by returning a level of control to the consumer. For instance, in 2019, Apple updated its privacy website with a revised declaration of the company's position on privacy.[1] This included a new section titled '*Control*' with a subsection titled '*Take Charge of Your Data*' which outlines:

> To give you more control over your personal information, we are rolling out a set of dedicated privacy management tools.

This relinquishing of control by an organization is enshrined in the concept of information stewardship. Information stewardship implies that no matter what an organization does with stakeholders' information (for example, selling it to third parties) the organization always remains responsible and retains oversight of the processing of that information (Rosenbaum 2010). Apple offer opportunities to their consumers that return a level of transparency and control to the consumer—by providing clear and simple explanations of what privacy options are available and how to implement them—they provide opportunities to their consumers that score highly on the Fair Information Practice Principles (FIPPs) scales. The principles of FIPPs are transparency, preference, purpose, minimization, limitation, quality, integrity, security, and accountability (DHS 2008, p. 4). Information privacy and FIPPs have been linked to procedural justice (Culnan and Bies 2003), where an organization's control over personal information is noted as 'just' when the information owner is vested with the key principles of FIPPs (Culnan and Bies 2003; Greenaway et al. 2015). Organizations demonstrating behaviors that exceed compliance obligations to protect information entrusted to them, view themselves as information stewards rather than information owners and

---

[1] https://www.macstories.net/news/apple-reveals-major-update-to-its-privacy-webpage/

demonstrate a 'culture of caring' with regard to privacy (Accenture and Ponemon 2015). These types of organizations have also been found to experience less privacy incidents (Accenture and Ponemon 2015).

5.3 Privacy and Cloud Computing

Although cloud computing enables reduced start-up costs, reduced operating costs and increased agility, its architectural features also raise various privacy concerns which are shared across several key stakeholders (Takabi et al. 2010) such as the CSP, the CSC, the consumer/data subject and application developers. Organizations are increasingly concerned about the risks of storing their data and applications on systems that reside outside of their on-premise data centres (Chen et al. 2010). ENISA (2009) define these risks as:

- privacy risks for the CSC e.g. non-compliance to enterprise policies and legislation, loss of reputation and credibility and being forced or persuaded to be tracked, or give personal information against their will (e.g. by governments)
- privacy risks for implementers of cloud platforms e.g. exposure of sensitive information stored on the platforms (potentially for fraudulent purposes), legal liability, loss of reputation and credibility, lack of user trust and take-up
- privacy risks for providers of applications on top of cloud platforms e.g. legal noncompliance, loss of reputation, 'function creep' using the personal information stored on the cloud, i.e. it might be used for purposes other than the original cloud intention
- privacy risks for the data subject e.g. exposure of personal information

However, unlike the traditional on-premise model of computing, responsibility for these privacy risks in the cloud computing model is shared between the CSC and the CSP, and the balance between the two responsibilities changes between cloud service models. This relationship is known as the shared responsibility model, and it is the basis for how modern cloud security and privacy operates with each service model differing in the amount of control offered to the CSC (Tripwire 2018). Haeberlen (2010) describes how different cloud service models affect the ways that privacy responsibilities are shared between CSPs and CSCs:

- IaaS—the CSP is responsible for the implementation and management of privacy controls only within the physical infrastructure. The CSC is responsible for all other aspects of privacy.
- PaaS—the CSP is responsible for IaaS. However, CSCs and CSPs are jointly responsible for ensuring appropriate privacy controls are implemented within the applications deployed on the PaaS environment.
- SaaS—the CSC has limited control over privacy and security. CSCs will generally maintain responsibility for managing identity and access management controls to ensure minimum permissions are assigned to roles. The CSP is responsible for ensuring all other privacy controls are in place.

As a CSC moves from on-premise models to cloud service models, they lose control over their data, including control over the privacy of that data. Although the shared responsibilities model assumes a level of transparency and control for the CSC, CSPs have traditionally lacked transparency regarding their privacy policies, strategy, service, thresholds etc. making it difficult for CSCs to objectively perform evaluations and risk assessments for a CSP service (Cruzes and Jaatun 2015). The implementation of ethical principles such as those in FIPPs not only mitigate the key privacy risks associated with cloud computing (Pearson 2009) but also offer the CSP an opportunity to rebalance the control-justice equilibrium for the CSC.

Many CSPs including Amazon, Microsoft, and IBM do offer simple breakdowns of performance metrics and responsibilities etc. However, some CSPs (e.g. SAAS model CSPs such as Salesforce or Workday) do not clearly define these sufficiently (Prüfer 2018). In a recent survey of IT decision makers (Netapp 2016) 35% believed responsibility for data sits with the CSPs, while 3% did not know who would be responsible. GDPR is very clear that responsibility for personal data lies firmly with the data controller (GDPR, Article 24). Under GDPR, the CSC is responsible (regardless of cloud computing model) for ensuring their own compliance requirements are handled effectively by the CSP and ensuring these requirements are adequately reflected in legally binding contractual agreements (called Data Processing Agreements, or DPAs) with the CSP. However, if a CSP is not transparent for instance, about which core IT services they themselves outsource to sub-processors, the CSC is unable to properly evaluate risks. In some cases, it may be difficult for the CSC (in its role as data controller) to assess the adequacy of the CSP's data handling practices to ensure that their data is handled in a lawful way. This problem is exacerbated in cases of multiple transfers of data e.g. between federated clouds. This lack of transparency is linked to decreased levels of trust in the CSP, is a key barrier for the adoption of cloud services (Del Alamo et al. 2015) and is associated with lack of accountability (Pearson 2009; Haeberlen 2010).

Finally, CSCs often overlook the on-premise privacy measures that traditional applications rely on such as on-premise firewall configurations that block logins from specific locations (such as embargoed countries), intrusion prevention systems, behavior analytics platforms (detecting insider threats), log management and alerting solutions (Oracle 2017). Since these and other measures protect privacy in applications in the enterprise campus, they are often taken for granted in the context of any one particular application and the responsibility for their installation and upkeep falls squarely on the CSC (Oracle 2017).

Alongside issues of transparency, responsibility and accountability, Abed and Chavan (2019) highlight a number of universal privacy issues facing CSCs, when considering cloud computing adoption, namely; the institutional obligation for disclosure (to governments), breach and incident disclosure, data accessibility and retention, and physical storage location):

- Institutional obligation for disclosure to Governments: Studies have shown that CSCs considering cloud adoption are concerned that data outsourced to the CSP can be accessed by others, notably public authorities with legitimate or illegitimate objectives as well as legal and illegal private actors (August et al. 2014). For instance in 2018, the Clarifying Lawful Overseas Use of Data Act (CLOUD Act, House of Representatives Bill 4943, 2018) enabled law enforcement agencies to access data processed by US-based companies, regardless of whether the servers were located in the US.
- Breach and incident disclosure: GDPR mandates that a privacy incident/breach be reported within 72 hours of its discovery. However, when the privacy incident/breach occurs in the CSP environment, it is very difficult for the CSC to have transparent discovery and disclosure arrangements in place. The breach disclosure requirements are defined under Article 28 of the GDPR and need to be incorporated into the DPA with the CSP. This contract also needs to clearly define the balance of liability in the event of a data breach.

- Data accessibility and retention: Guaranteeing availability of cloud data when migrating from one CSP to another has become a primary concern for CSCs (Xue et al. 2017). In the CSP environment, personal data accessibility, retention and deletion become fundamental requirements to prevent CSCs being 'locked in' to a given CSP.
- Physical Storage Location: CSPs may store data in multiple geographically dispersed jurisdictional locations. Often this presents a challenge to identify which legislation takes precedence where laws conflict, particularly where there is conflict between the laws applying to the physical location of a CSP data centre and the physical location of the CSC (Abed and Chavan 2019).

5.4   The Privacy Orientation Framework

In the early years of privacy, organizations understood their responsibilities toward privacy to be legal and financial responsibilities. In the 70's discretionary frameworks such as FIPPs emerged—combining privacy standards with due process, consumer rights, and equality protections (Westin 2003). From the turn of the century Westin (2003) suggests that privacy became a first-level social and political issue in response to 9/11, the Internet, the cell phone, the human genome project, data mining, automation of government public records amongst others. In the last decade organizations began to respond to these fundamental changes to concern for privacy, with initiatives that exceeded their legal, financial and ethical responsibilities—ranging from privacy-by-design standards, developing open privacy standards, to collaborating with privacy advocacy groups. Given their association with justice and improved privacy protection, our framework was concerned with those privacy behaviors exceeding legislation. Much of the literature investigating privacy beyond legislation (Pollach 2011; Allen and Peloza 2015) explored privacy as a Corporate Social Responsibility (CSR). McWilliams and Siegel (2001) define CSR as those actions that appear to further some social good, beyond the interests of the organization and 'beyond that which is required by law'.

Carroll's model of CSR is the most commonly known model for CSR (Visser 2006). Carroll (1979) identified four pillars of CSR (financial responsibilities, legal responsibilities, ethical responsibilities, and philanthropic responsibilities). Privacy is a financial responsibility as organizations can be fined significant sums of money for non-compliance. Privacy is also a legal responsibility as privacy legislation (e.g. the California Consumer Privacy Act (CCPA) or the GDPR) mandates strict governance over the processing of personal data. Carroll (1998) suggested that privacy not only is a legal and financial responsibility but is also an ethical responsibility, as legislation lags behind ethics, and morality comes into play. Privacy also meets Mason's (1995) test of what constitutes an ethical problem i.e. whenever one party in pursuit of its goals engages in behavior that materially affects the ability of another party to pursue its goals. In 2018, the then EU Data Protection Supervisor (Giovanni Buttarelli) argued that in order to address this ethical component of privacy, organizations should not overly rely on bare compliance with the letter of law, and should adopt a 'duty of care' for consumer data (Buttarelli 2018). CSR activities are most appropriate where existing legislation requires compliance with the spirit as well as the letter of the law and where the organization can fool stakeholders through superior knowledge (Mintzberg 1983).

Finally, privacy can be described as a philanthropic responsibility, where an organization's privacy behaviors demonstrate a 'duty of care' (Buttarelli 2018) towards data owners and society that exceeds compliance. Philanthropy presents itself in many forms e.g. cash contributions or employee commitments (Smith 1994) and is often explained by social exchange theory (Emerson 1962) where corporations use philanthropy to expand the scope of their business initiatives, influence governments, and position themselves as influential leaders (Jung et al. 2016). Both Husted (2003) and Stannard-Stockton (2011) suggest three classifications of philanthropy; (1) "check-book philanthropy" i.e. making cash contributions to a cause; (2) in-house projects and philanthropic investments; (3) strategic philanthropic collaboration between organizations and non-corporate partners. True philanthropy matches the resources of the giver with the needs of the recipient through a socially beneficial relation that is mobilized and governed by a force of morally armed entreaty (Schervish 1998).

As we could find no reference in the literature to philanthropic privacy, based on definitions of CSR from McWilliams and Siegel (2001) we describe it as: '*Any privacy behavior(s) exceeding legislative requirements, that furthers privacy as a societal good, beyond the interests of the organization*'. Philanthropic privacy behaviors would therefore include behaviors such as advising on government policy, developing open privacy standards or tools, exceeding privacy laws for employees, even lobbying for strengthened privacy on behalf of the consumer or society. Whilst traditional lobbying often aims at shaping rules for privacy in the best interest of the organization, some organizations may try to shape the rules in a way that favors society's needs for privacy. The inclusion of the philanthropic dimension is important, as organizations taking privacy actions beyond their corporate obligations have been found to experience less privacy incidents (Accenture and Ponemon 2015). It therefore seems reasonable to suggest that the philanthropic privacy dimension may help identify privacy behaviors that could help strengthen privacy protection effectiveness.

Although privacy behaviors 'beyond that which is required by law' may demonstrate the differentiating behaviors of justice that this chapter is keen to explore, it is important to highlight that this concept of 'beyond that which is required by law' differs from one jurisdiction to another. For instance, the GDPR mandates that organizations provide transparency, choice, purpose and notice to the data subject. Many other jurisdictions, particular large sectors of the US, do not mandate the provision of these privacy behaviors through regulation but through industry self-regulation and discretionary codes such as FIPPs. Thus, choice for example, would be considered a legal requirement within GDPR, and in certain US contexts an ethical requirement. This is an important consideration for any future application of our framework, as an evaluation of the local privacy landscape for a given organization would first be required, in order to determine their legal minimums.

5.4.1   Framework Dimensions

Greenaway et al. (2015), combine control and justice theory to form a privacy orientation which they call Company Information Privacy Orientation (CIPO). Their research forms the starting point for the construction of the privacy orientation framework we develop in this chapter. Greenaway et al. (2015) suggest that an organization's privacy orientation is founded on three dimensions namely: information management (i.e. how an organization uses information for profit), legal and ethical dimensions. These three dimensions are underpinned by several theoretical components: stakeholder theory, stockholder theory, social contract theory, information processing theory and cognitive categorization theory (Greenaway et al. 2015). Through triangulation of their three CIPO dimensions with the four pillars of CSR from Carroll's (1979) CSR pyramid of responsibilities (economic, legal, ethical, and philanthropic), we extend the CIPO framework with the addition of the philanthropic dimension. Triangulation enables a systematic search for convergence among sources of information to form new or revised themes/categories representing the state of knowledge for a given subject (Denzin 1978; Yardley 2008). See Table 5.1 for the Privacy Orientation Framework

Table 5.1 Privacy orientation framework dimensions including philanthropy

Dimensions	Definition	Theory base	Attribute range	Behavior definition
Philanthropic responsibilities	The firms contributions to society	Social exchange theory (Emerson 1962). CSR theories (Carroll 1998). Triple bottom line theory of CSR (Savitz 2013).	Checkbook philanthropy	Paying for charitable events or activities that promote privacy, outside of the organisation.
			Inhouse Projects	Participation in groups/activities aimed at supporting privacy for employees or extended supply chain.
			Strategic Collaboration	Participating in groups/activities aimed at supporting privacy for consumers, society, humanity.
Ethical responsibilities	The firms' obligations to its consumers	Ethical theories: Smith and Hasnas (1999). Social contract theory (Beauchamp and Bowie 1993; Donaldson and Dunfee 1999).	Stockholder theory	Consumers are obligated to protect themselves in transaction with the firm
			Stakeholder theory	Firm and consumer share responsibility for consumer wellbeing
			Social contract theory	Firm obligated to promote consumer wellbeing.

(*continued*)

Table 5.1 (continued)

Dimensions	Definition	Theory base	Attribute range	Behavior definition
Information management/ financial responsibilities	The firm's purpose for gathering and using information— how it intends to make profit	Information processing theory (Tushman and Nadler 1978) as used by Brohman et al. (2003)	Transaction focused	Consumer focused transaction data collected for short term efficiency
			Data focused	Consumer data collected from various sources to improve profitability
			Inference focused	Consumer profiles created from historic/ preference data: cross-sell & up-sell
			Advice focussed	Consumer profiles created from historic, preference, aspirational data to create profitable sustainable relationships
Legal and risk responsibilities	The firm's view of the impact that privacy laws have on its ability to carry out business	Cognitive categorisation theory (Rosch 1978) as used by Jackson and Dutton (1988)	No threat no opportunity	Firms lack awareness or concern for privacy laws
			Threat	Privacy laws viewed as constraint to maximize shareholder value
			Threat and Opportunity	Privacy laws viewed as required to ensure ongoing transactions with consumers
			Opportunity	Privacy laws embraced as strategic differentiation

Adapted and extended from Greenaway et al. (2015)

dimensions, adapted from the Greenaway et al. (2015) CIPO framework, extended to include details for the Philanthropic dimension.

A CSP may demonstrate a combination of both control-based behaviors and justice-based behaviors in varying levels. We therefore position control and justice as two continua, where four privacy orientations emerge, namely; 'low-justice, low-control' orientations (we call CSPs in this orientation 'Compliers'); 'low-justice, high-control' (we call CSPs in this orientation 'Integrators'); 'high-justice, high-control' (we call CSPs in this orientation 'Citizens'); and 'high-justice, low-control' (we call CSPs in this orientation 'Warriors'). We summarize our privacy orientation framework below in Fig. 5.1.

5.4.1.1 Compliers

Complier CSPs focus on privacy risk mitigation. CSPs in this orientation demonstrate privacy behaviors that measure low on justice and low on control. This orientation sets compliance as the goal, with privacy governance assigned to functional management. Their privacy policies and

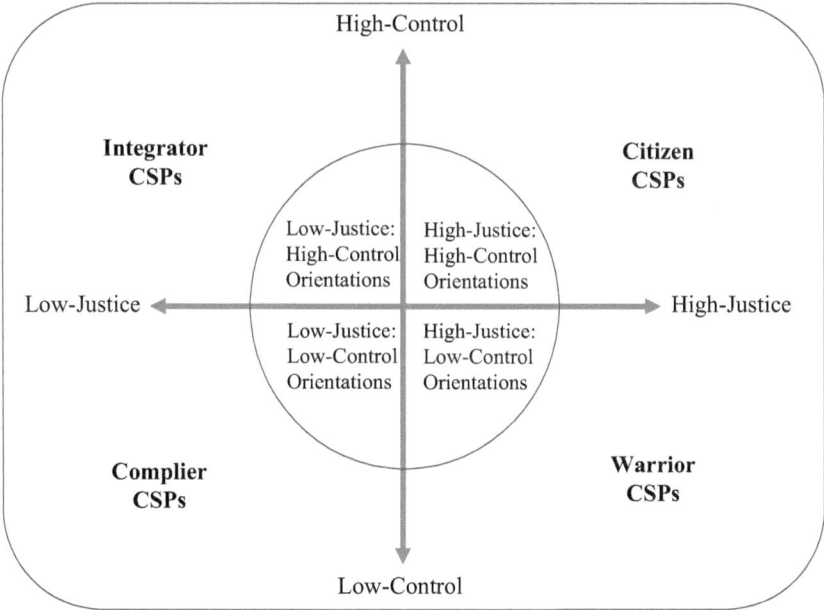

Fig. 5.1 Privacy orientations framework

practices are often centred on compliance with laws and industry standards. These organizations report little more than privacy metrics mandated by privacy. They may undertake lobbying towards privacy but only where beneficial to the organization. They have no real philanthropic perspective with regard to privacy. Compliers tend to provide rudimentary transactional services (for instance Cloudways and Digital Ocean). They will typically offer security and privacy features at an additional cost, or like Amazon AWS, allow the CSC to reconfigure private buckets to be non-private. Risk-Managers rarely exceed legislative minimums.

In 2019, Capital One (using AWS as their CSP) suffered a breach impacting 100 m customers' financial data including social security numbers (Business Insider 2019). The AWS server used by Capital One was vulnerable to a well-known attack (called server-side request forgery [SSRF]). AWS denied liability (as their contracts indemnified them) however—without being legally required to do so, AWS's largest competitors (Google and Microsoft) had already addressed the threat of SSRF attacks two years previously.[2]

5.4.1.2 Integrators

Integrator CSPs offer robust compliance to privacy legislation whilst integrating society's increasing expectations to address privacy as a social responsibility. CSPs in this privacy orientation demonstrate privacy behaviors that measure low on justice and high on control. Relinquishing control (by offering increased justice behaviors) to their CSCs is not important to these organizations, as they require widespread access to data and systems in order to minimize costs and maximize profit. CSPs in this orientation may actively reflect on ways they can use social issues such as privacy to gain competitive advantage. For instance, privacy rights for employees, their families, and local communities may be supported. The CSPs objective in this orientation is to mitigate the erosion of economic value in the medium term and to achieve longer-term gains by integrating responsible privacy practices into their daily operations (Zadek 2004). These CSPs not only want to monetize their cloud service, but they also want to maximize data value. Although CSPs in this orientation may appear to demonstrate philanthropic privacy—for example by producing white papers, open

[2] https://www.wyden.senate.gov/imo/media/doc/102419%20Wyden%20Warren%20Letter%20to%20FTC%20RE%20Amazon%20Capital%20One%20Hack.pdf

standards, open tools etc. this philanthropy towards privacy is not always a 'felt' value.

For instance, Google Cloud's published privacy commitments (Google 2020) state that "*we believe that trust is created through transparency, and we want to be transparent about our commitments and what you can expect when it comes to our shared responsibility for protecting and managing your data in the cloud*". Yet in 2019, Google were fined €50m by the French data protection authority for lack of transparency in their privacy policies (CNIL 2019).

5.4.1.3 Citizens

Citizen CSPs assume a citizenship role, leading privacy issues and transforming their business models to achieve this objective. Citizen CSPs demonstrate privacy behaviors that measure high on justice and high on control. They openly acknowledge their new roles and responsibilities towards society, and recognize how responsibilities towards private, public, and social sectors have become interdependent (Latapi-Agudelo et al. 2019). The Citizen orientation addresses privacy strategically, where privacy programs are focused on building and maintaining sustainable relationships, and often exceed legislative obligations. These organizations incorporate privacy as a value held strongly by the organization, associating their privacy behaviors with other strong values such as trust and integrity (Mirvis and Googins 2006). CSPs in the Citizen privacy orientation broaden their agenda by expanding their privacy concerns and deepen the involvement of top management in the leadership addressing privacy issues. Organizations form long-term alliance and partnerships with stakeholders in order to drive change in privacy issues. Caldwell et al. (2010) refer to this leadership as a 'stewardship role' which generates commitment from other stakeholders and organizations in order to drive change in several key privacy issues. CSPs in this orientation consider privacy to be a compliance obligation however they also consider privacy to be strategic, where privacy is used to build and maintain sustainable relationships, and often exceeds legal minimums—reflecting philanthropic privacy.

Microsoft for instance established an online trust centre to respond to queries on privacy for all its products, particularly on Office 365. They also provide tools to simplify responding to Subject Access Requests mandated by GDPR (Microsoft 2018a). Microsoft has also extended the rights available to Europeans under the GDPR, to all its consumers, noting that "*GDPR establishes important principles that are relevant globally*"

(Microsoft 2018b). IBM, as another example, were among the first companies to sign the EU Data Protection Code of Conduct for Cloud Service Providers (IBM 2017).

5.4.1.4   Warriors

Warrior CSPs highlight privacy as a societal issue and may even revolt against laws in order to support it. CSPs who are positioned in this orientation demonstrate privacy behaviors that measure high on justice, however unlike those of the Citizen, their behaviors may also measure low on control. In terms of philanthropic privacy, Warrior CSPs may lobby government for increased privacy for individuals, knowing it will increase organizational costs, reduce shareholder returns and/or marginalize certain stakeholders. They may stake claims to privacy, to such a degree that they will breach laws if they feel that those laws compromise social or democratic freedoms.

Apple for example, in 2015 and 2016, refused to comply with, and challenged, at least 12 orders issued by the FBI compelling Apple to enable decryption of phones involved in criminal investigations and prosecutions. It could be argued that Apple monetize 'privacy' as part of their brand, however in the US they did not garner widespread support for their decision and at that time 51% of American smartphone users were against Apple's decision, while only 38% supported Apples stance (Pew 2016). Warriors may demonstrate less control than those in the Citizen orientation, however in its place they introduce strong privacy governance and accountability structures. In Apple for instance, any collection of customer data requires sign-off from a committee of three "privacy czars" and a top executive (Reuters 2016).

Cisco, as another example, recognized that their systems could be used in a manner contrary to their values and state *"Cisco technologies are used by government agencies to promote public safety, but the same technology can be used for surveillance that would violate individuals' privacy"*. (Cisco 2018c). Cisco issued a position statement opposing government backdoors and opposing attempts to prohibit public disclosure about new surveillance capabilities demanded by governments (Cisco 2018b). Cisco therefore retain a high level of control to ensure that their product cannot be misused but also balance this control with a commitment to *"build our products on the open, global standards we believe are critical to overcoming censorship, protecting privacy, and keeping the world connected"* (Cisco 2018c). Cisco also undertakes several philanthropic privacy activities such as the Cisco Privacy Maturity Benchmark Study and sponsoring the National Cybersecurity Alliance's Data Privacy Day (Cisco 2018c).

5.5   Conclusion

Researchers, practitioners, and policymakers need a better understanding of how and why organizations differ in their treatment of consumer privacy. Belanger and Xu (2015) advocate that privacy researchers conduct qualitative interpretive research about relationships between privacy antecedents and outcomes. Using this framework, such research could be achieved by categorizing and weighing levels of control and justice demonstrated by a CSP's privacy behaviors and positioning them as one of the four orientations in the framework. The framework also presents opportunities for such future research to empirically examine the effectiveness of one orientation over another, or the potential impact of an organization's orientation on other variables such as privacy protection effectiveness, cybersecurity behaviors, or privacy strategies. The framework can also be applied in different contexts—comparing privacy orientations for CSPs in different countries, of different sizes, or with different market dependencies on data.

Being able to position a CSP's privacy orientation allows the CSP to establish baselines within their industry and to determine if certain orientations provide more robust security and privacy protection, where privacy incidents may be reduced. At the same time, the framework challenges CSPs to articulate their ethical, financial, legal, and philanthropic strategies. Moreover, identifying their privacy orientation will assist CSPs to better align their actual privacy programs with the CSCs top concerns (Chan 2003). If CSPs can use their privacy orientation to enhance and make more effective their privacy provisions, then the wider stakeholder community, particularly the CSC, and ultimately the data subject, may benefit. We hope the framework provides a helpful foundation for future theoretical and empirical work.

Acknowledgements This chapter and the work described therein was funded by the Irish Research Council's Employment Based Post-graduate Scholarship programme.

REFERENCES

Abed, Y., & Chavan, M. (2019). The Challenges of Institutional Distance: Data Privacy Issues in Cloud Computing. *Science, Technology and Society, 24*(1), 161–181. https://doi.org/10.1177/0971721818806088.

Accenture and Ponemon. (2015). How Global Organizations Approach the Challenge of Protecting Personal Data. Retrieved June 2020, from http://www.ponemon.org/local/upload/file/ATC_DPP%20report_FINAL.pdf

Acquisti, G. (2008). Identity Management, Privacy, and Price Discrimination. *IEEE Security and Privacy, 6*(1), 46–50. https://doi.org/10.1109/MSP.2008.35.

Allan, L., & Tyler, T. (1988). *The Social Psychology of Procedural Justice.* Plenum Press New York.

Allen, A., & Peloza, J. (2015). Someone to Watch Over Me: The Integration of Privacy and Corporate Social Responsibility. *Business Horizons, 58*, 635–642. https://doi.org/10.1016/j.bushor.2015.06.007.

Ashworth, L., & Free, C. (2006). Marketing Dataveillance and Digital Privacy: Using Theories of Justice to Understand Consumers Online Privacy Concerns. *Journal of Business Ethics, 67*, 107–123.

August, T., Niculescu, M., & Shin, H. (2014). Cloud Implications on Software Network Structure and Security Risks. *Information Systems Research, 25*(3), 489–510.

Beauchamp, T., & Bowie, N. (1993). *Ethical Theory and Business.* Chambersburg, PA: Prentice Hall.

Belanger, F., & Crossler, R. (2011). Privacy in the Digital Age: A Review of Information Privacy Research in Information Systems. *MIS Quarterly, 35*, 1017–1041.

Belanger, F., Hiller, J., & Smith, W. (2002). Trustworthiness in Electronic Commerce: The Role of Privacy, Security, and Site Attributes. *Journal of Strategic Information Systems, 11*, 245–270.

Belanger, F., & Xu, H. (2015). The Role of Information Systems Research in Shaping the Future of Privacy. *Information System Journal, 25*, 573–578.

Brohman, M., Watson, R., Piccoli, G., & Parasurama, A. (2003). Data Completeness: a Key to Effective Net-Based Customer Service Systems. *Communications of the ACM, 46*, 47–51.

Business Insider. (2019). To Prevent Disasters like the Capital One Hack from Happening Again, Experts Say Amazon Web Services Could Do More to Protect Customers from Themselves. Retrieved from May 2020, from https://www.businessinsider.com/capital-one-hack-amazon-aws-breach-security-analysts-2019-8?r=US&IR=T

Buttarelli, G. (2018). Interview with the European Data Protection Supervisor Giovanni Buttarelli: 'The GDPR is a Radical Update to the Rulebook for the

Digital Age'. Retrieved September 30, 2018, from https://irishtechnews.ie/interview-with-the-european-data-protection-supervisor-giovanni-buttarelli-the-gdpr-is-a-radical-update-of-the-rule-book-for-the-digital-age/

Caldwell, C., Hayes, L., & Long, D. (2010). Leadership, Trustworthiness, and Ethical Stewardship. *Journal of Business Ethics, 96*(4), 497–512.

Carroll, A. (1979). A Three Dimensional Conceptual Model of Corporate Performance. *Academy of Management Review, 4*, 497–505.

Carroll, A. (1998). The Four Faces of Corporate Citizenship. *Business and Society Review, 100*, 1–7.

Caudill, E., & Murphy, P. (2000). Consumer Online Privacy: Legal and Ethical Issues. *Journal of Public Policy and Marketing, 19*(1), 7–19.

Chan, Y. (2003). Competing Through Information Privacy. In J. N. Luftman (Ed.), *Competing in the Information Age: Align in the Sand* (2nd ed., pp. 350–361). New York: Oxford University Press.

Chen, H., Chiang, R., & Storey, V. (2012). Business Intelligence and Analytics: From Big Data to Big Impact. *MIS Quarterly, 36*, 1165–1188.

Chen, Y., Paxson, V., & Katz, R. (2010). *What's New About Cloud Computing Security?* Tech. Report UCB/EECS-2010-5, EECS Department, University of California, Berkeley. Retrieved June 2020, from www.eecs.berkeley.edu/Pubs/TechRpts/2010/EECS-2010-5.html.

Cisco. (2018a). Global Cloud Index White Paper. Retrieved June 2020, from https://www.cisco.com/c/en/us/solutions/collateral/service-provider/global-cloud-index-gci/white-paper-c11-738085.html

Cisco. (2018b). Cisco Position Statement on Human Rights and Privacy. Retrieved June 2020, from https://www.cisco.com/c/dam/assets/csr/pdf/Human-Rights-Position-Statements-2018.pdf

Cisco. (2018c). Cisco Corporate Social Responsibility Report. Retrieved May 2020, from https://www.cisco.com/c/dam/assets/csr/pdf/CSR-Report-2018.pdf

CNIL. (2019). Commission Nationale D'Information Liberation. (The French Data Protection Supervisory Authority). Retrieved from https://www.cnil.fr/en/cnils-restricted-committee-imposes-financial-penalty-50-million-euros-against-google-llc

Consumers International. (2019). The State of Data Protection Rules Around the World: A Briefing for Consumer Organizations. Retrieved May 2020, from https://www.consumersinternational.org/media/155133/gdpr-briefing.pdf

Cruzes, D., & Jaatun, M. (2015). *Cloud Provider Transparency—A View from Cloud Customers*. 5th International Conference on Cloud Computing and Services Science.

CSA. (2009). Cloud Security Alliance. Security Guidance for Critical Areas of Focus in Cloud Computing. Retrieved May 2020, from https://cloudsecurity-alliance.org/csaguide.pdf

Culnan, M., & Armstrong, P. (1999). Information Privacy Concerns, Procedural Fairness, and Impersonal Trust: An Empirical Investigation. *Organizational Science, 10*, 104–115. https://doi.org/10.1287/orsc.10.1.104.

Culnan, M., & Bies, R. (2003). Consumer Privacy: Balancing Economic and Justice Considerations. *Journal of Social Issues, 59*, 323–342. https://doi.org/10.1111/1540-4560.00067.

Culnan, M., & Williams, C. (2009). How Ethics Can Enhance Organizational Privacy: Lessons from the ChoicePoint and TJX Data Breaches. *MIS Quarterly, 33*, 673–687.

Davis, K., Frederick, W., & Blomstrom, R. (1980). *Business and Society* (2nd ed.). New York, NY: McGraw-Hill.

Del Alamo, J., Trapero, R., Martín, Y., & Yelmo, J. (2015). Assessing Privacy Capabilities of Cloud Service Providers. *IEEE Latin America Transactions, 13*(11), 3634–3641.

Denzin, N. (1978). *Sociological Methods Sourcebook* (2nd ed.). NY: McGraw Hill.

DHS. (2008). *The Fair Information Practice Principles: Framework for Privacy Policy at the Department of Homeland Security.* Memorandum Number: 2008-01. US Department of Homeland Security Privacy Office Publications. Retrieved April 2020, from https://www.dhs.gov/xlibrary/assets/privacy/privacy_policyguide_2008-01.pdf

DHS. (2017). *Handbook for Safeguarding Sensitive Personal Information.* Privacy Policy Directive 047-01-007, Revision 3. US Department of Homeland Security Privacy Office Publications. Retrieved April 2020, from https://www.dhs.gov/sites/default/files/publications/dhs%20policy%20directive%20047-01-007%20handbook%20for%20safeguarding%20sensitive%20PII%2012-4-2017.pdf

Dinev, T. (2014). Why Would We Care about Privacy? *European Journal of Information Systems, 23*, 97–102.

Donaldson, T., & Dunfee, T. (1999). *Ties That Bind: A Social Contracts Approach to Business Ethics.* Boston, MA: Harvard Business School Press.

Emerson, R. (1962). Power-Dependence Relations. *American Sociological Review, 27*(1), 31–41.

ENISA. (2009). Cloud Computing: Benefits, Risks and Recommendations for Information Security, by Catteddu, D. and Hogben, G. Retrieved April 2020, from www.enisa.europa.eu/act/rm/files/deliverables/cloud-computing-risk-assessment/at_download/fullReport

Forbes. (2017). Global Cloud Spending Predicted to Reach $390b by 2020. Retrieved May 2020, from https://www.forbes.com/sites/louiscolumbus/2017/02/11/global-cloud-spending-predicted-to-reach-390b-by-2020/#125897191085

Fortune. (2019). Capital One's Data Breach Could Cost the Company Up to $500 Million. Retrieved May 2020, from https://fortune.com/2019/07/31/capital-one-data-breach-2019-paige-thompson-settlement/

Fried, C. (1984). *Philosophical Dimensions of Privacy* (E. D. Schoeman, Ed., pp. 203–222). New York: Cambridge University Press.

Google. (2020). Privacy on GoogleCloud. Retrieved May 2020, from https://cloud.google.com/security/privacy

Greenaway, K., Chan, Y., & Crossler, R. (2015). Company Information Privacy Orientation. A Conceptual Framework. *Information Systems Journal, 25*, 579–606.

Haeberlen, A. (2010). A Case for the Accountable Cloud. *SIGOPS Operating Systems Review, 44*(2), 52–57.

Husted, B. (2003). Governance Choices for Corporate Social Responsibility: To Contribute, Collaborate or Internalize? *Long Range Planning, 36*(5), 481–498.

IBM. (2017). IBM Among the First Companies to Sign EU Data Protection Code of Conduct for Cloud Service Providers. BlogPost by Chrstine Capella, CPO IBM. Retrieved May 2020, from https://www.ibm.com/blogs/policy/eu-cloud-code-of-conduct/.

Jackson, S., & Dutton, J. (1988). Discerning Threats and Opportunities. *Administrative Science Quarterly, 33*(1), 370–387.

Johnson, D. (2009). *Computer Ethics* (3rd ed.). Upper Saddle River, NJ: Pearson Education Inc. Retrieved May 2020, from https://www.ibm.com/blogs/policy/eu-cloud-code-of-conduct/

Jung, T., Phllips, S., & Harrow, J. (2016). *The Routledge Companion to Philanthropy*. London, UK: Routledge Press.

Laczniak, G., & Murphy, P. (1993). *Ethical Marketing Decisions: The Higher Road*. Boston: Allyn & Bacon.

Latapi-Agudelo, M., Johannsdottir, L., & Davidsdottir, B. (2019). A Literature Review of the History and Evolution of Corporate Social Responsibility. *International Journal of Corporate Social Responsibility, 4*(1), 1–23.

Martin, K. (2012). Diminished or Just Different? A Factorial Vignette Study of Privacy as a Social Contract. *Journal of Business Ethics, 11*(4), 519–539.

Martin, K. (2016). Understanding privacy online: Development of a Social Contract Approach to Privacy. *Journal of Business Ethics, 137*(3), 551–569.

Mason, R. (1995). Apply Ethics to Information Technology Issues. *Communications of the ACM, 38*, 55–57.

McWilliams, A., & Siegel, D. (2001). Corporate Social Responsibility: A Theory of the Firm Perspective. *Academy of Management Review, 26*, 117–127.

Mell, P., & Grance, T. (2011). *The NIST Definition of Cloud Computing*. US National Institute of Science and Technology. Retrieved November 2019, from http://csrc.nist.gov/publications/nistpubs/800-145/SP800-145.pdf

Microsoft. (2018a). Microsoft Trust Centre GDPR Compliance. Retrieved from https://www.microsoft.com/en-us/TrustCenter/CloudServices/office365/GDPR

Microsoft. (2018b). Microsoft's Commitment to GDPR, Privacy and Putting Customers in Control of Their Own Data. https://blogs.microsoft.com/on-the-issues/2018/05/21/microsofts-commitment-to-gdpr-privacy-and-putting-customers-in-control-of-their-own-data/

Mintzberg, H. (1983). *Power In and Around Organizations*. Englewood Cliffs, NJ: Prentice-Hall.

Mirvis, P., & Googins, B. (2006). Stages of Corporate Citizenship. *California Management Review, 48*(2), 104–126.

Moor, J. (1997). Towards a Theory of Privacy in the Information Age. *Computers and Society, 27*, 27–32.

Mulligan, D., Koopman, C., & Doty, N. (2016). Privacy is an Essentially Contested Concept: A Multi-dimensional Analytic for Mapping Privacy. *Philosophical Transactions. Series A, Mathematical, Physical, and Engineering Sciences, 374*(2083), 20160118.

Netapp. (2016). Retrieved November 2019, from https://www.netapp.co.uk/company/news/press-releases/news-rel-20160712-103195.aspx

Oracle. (2017). White Paper: "Making Sense of the Shared Responsibility Model". Retrieved November 2019, from http://www.oracle.com/us/solutions/cloud/platform-as-a-service/shared-responsibility-model-wp-3497462.pdf

Pearson, S. (2009). *Taking Account of Privacy When Designing Cloud Computing Services*. Proceedings of the 2009 ICSE Workshop on Software Engineering Challenges of Cloud Computing, pp. 44–52.

Pew. (2016). *More Support for Justice Department Than for Apple in Dispute Over Unlocking iPhone*. Pew Research Center for the People and the Press.

Pollach, I. (2011). Online Privacy as a Corporate Social Responsibility: An Empirical Study. *Business Ethics: A European Review, 20*, 88–103.

Prüfer, J. (2018). Trusting Privacy in the Cloud. *Information Economics and Policy, 45*, 52.

Ren, K., Wang, C., & Wang, Q. (2012). Security Challenges for the Public Cloud. *IEEE Internet Computing., 16*(1), 69–73.

Reuters. (2016). Apple Privacy Czars Grapple With Internal Conflict Over User Data. Retrieved October 2019, from https://www.reuters.com/article/us-apple-encryption-privacy-insight-idUSKCN0WN0BO

Rosch, E. (1978). Principles of Categorization. In E. Rosch & B. Lloyd (Eds.), *Cognition and Categorization*. Lawrence Elbaum Associates. ISBN 0835734048, 9780835734042.

Rosenbaum, S. (2010). Data Governance and Stewardship: Designing Data Stewardship Entities and Advancing Data Access. *Health Services Research, 45*, 1442–1455.

Savitz, A. (2013). *The Triple Bottom Line: How Today's Best-Run Companies Are Achieving Economic, Social and Environmental Success—and How You Can Too*. San Francisco, CA: Jossey-Bass Press.

Schervish, P. (1998). Philanthropy. In R. Wuthnow (Ed.), *Encyclopaedia of Politics and Religions* (pp. 600–603). Washington, DC: Congressional Quarterly.

Smith, C. (1994). The New Corporate Philanthropy. *Harvard Business Review, 72*(3), 105–116.

Smith, H., Dinev, T., & Xu, H. (2011). Theory and Review of Information Privacy Research: An Interdisciplinary Review. *MIS Quarterly, 35,* 989–1015.

Smith, H., & Hasnas, J. (1999). Ethics and Information Systems: the Corporate Domain. MIS Quarterly, *23,* 109–127.

Stannard-Stockton, S. (2011). The Three Core Approaches to Effective Philanthropy. *Stanford Social Innovation Review.* Retrieved from https://ssir.org/articles/entry/the_three_core_approaches_to_effective_philanthropy#

Takabi, H., Joshil, J., & Ahn, G. (2010). *Security and Privacy Challenges in Cloud Computing Environments.* IEEE Computer and Reliability Societies.

Techcrunch. (2019). Data Management Giant Rubrik Leaked a Massive Database of Client Data. Retrieved May 2020, from https://techcrunch.com/2019/01/29/rubrik-data-leak/

Tripwire. (2018). Cloud Security Shared Responsibility Model Explained. Retrieved September 2019, from https://www.tripwire.com/state-of-security/security-data-protection/cyber-security/cloud-securitys-shared-responsibility-model-explained/

Tushman, M., & Nadler, D. (1978). Information Processing as an Integrating Concept in Organizational Design. *Academy of Management Review, 3,* 613–624.

Visser, W. (2006). Revisiting Carroll's CSR Pyramid: An African Perspective. In E. Pedersen & M. Huniche (Eds.), *Corporate Citizenship in Developing Countries* (pp. 29–56). Copenhagen Business School Press.

Westin, A. (2003). Social and Political Dimensions of Privacy. *Journal of Social Issues, 59*(2), 431–453.

Wright, S., & Xie, G. (2019). Perceived Privacy Violation: Exploring the Malleability of Privacy Expectations. *Journal of Business Ethics, 156*(1), 123–140. https://doi.org/10.1007/s10551-017-3553.

Xu, H., Tan, B., & Agarwal, R. (2012). Effects of Individual Self-Protection, Industry Self-Regulation, and Government Regulation on Privacy Concerns: A Study of Location-Based Services. *Information Systems Research, 23,* 1342–1363.

Xue, L., Ni, J., & Shen, J. (2017). Provable Data Transfer from Provable Data Possession and Deletion in Cloud Storage. *Journal of Computer Standards and Interfaces, 54*(1), 46–54.

Yardley, L. (2008). Demonstrating the Validity of Qualitative Research. *The Journal of Positive Psychology, 12,* 295–296.

Zadek, S. (2004). The Path to Corporate Responsibility. *Harvard Business Review, 82*(12), 125–132.

CHAPTER 6

Ethics and Cloud Computing

Brid Murphy and Marta Rocchi

Abstract While the benefits of cloud computing are widely acknowledged, it raises a range of ethical concerns. The extant cloud computing literature reports specific ethical perspectives on focussed topics in this domain, but does not explicitly refer to a specific ethical conception or reference point. This chapter provides an overview of ethics and ethical theories, which can be used to analyse the use of cloud technology and the complex multi-stakeholder structure of the industry. It is critical that cloud providers and users recognise that they effectively shape the morality of the cloud computing context through their interactions with other providers and users, and with the platform itself. Both stakeholder sets must be accountable for the possibilities offered by the technology. While pertinent regulation is continuously evolving, it is unlikely to advance at a

B. Murphy (✉)
DCU Business School, Irish Institute of Digital Business, Dublin City University, Dublin, Ireland
e-mail: brid.murphy@dcu.ie

M. Rocchi
DCU Business School, Irish Institute of Digital Business, Dublin City University, Dublin, Ireland
e-mail: marta.rocchi@dcu.ie

© The Author(s) 2021
T. Lynn et al. (eds.), *Data Privacy and Trust in Cloud Computing*,
Palgrave Studies in Digital Business & Enabling Technologies,
https://doi.org/10.1007/978-3-030-54660-1_6

similar rapid pace to that of innovation in the cloud computing industry. It is therefore essential that ethics is carefully considered to orient cloud computing towards the good of society.

Keywords Ethics • Cloud computing • Ethical analysis • Ethics of technology • Data ownership • Data privacy • Green computing • Organisational ethics • Personal ethics

6.1   INTRODUCTION

The many benefits and risks associated with the adoption of cloud computing technology have been widely reported by computer scientists. However, the same cannot be said in relation to ethical issues associated with the adoption of this technology: while voices caution that cloud technologies may traverse the boundaries of what is morally desirable for individuals, business, and society, few academics with ethics expertise have actually analysed the ethical issues associated with cloud computing. Contributions on ethical concerns connected with this new technology have been published mainly in computer science journals and have been discussed at conferences with a similar audience—sometimes privileging the explanation of technical aspects over the depth of the ethical analysis.

In light of this, the chapter aims to review selected ethical issues related to cloud computing, arising from the nature and use of the technology and the way in which the industry is structured, making reference to the existing literature on the topic. Section 6.2 provides an overview on ethics and some prominent ethical theories. It establishes the significance of ethical analysis when applied to a particular domain, such as cloud computing. The section concludes by questioning the depth of ethical analysis conducted in relation to cloud computing. It also details how the remainder of the chapter is organized, in accordance with recurring themes in existing literature; however, these do not take into explicit consideration particular schools of ethics when describing ethical issues associated with cloud computing, as the literature to date has failed to do so. Section 6.3 categorizes existing literature specifically related to cloud computing in accordance with three different and interrelated perspectives: it critically reviews ethical perspectives related to data, it then examines the ethical perspectives pertinent to providers of cloud computing services, and it also explores ethical issues related to the final users of cloud computing services. Each of these sub-sections presents literature within the context of ethical analysis as outlined in Sect. 6.2; it should however be noted that due to the limitations previously outlined on the depth of ethical analysis, the section is sometimes presented as a critical collection of existing contribution on the theme. The Conclusions section hosts final reflections and outlines avenues for further research; ultimately, it suggests that ethical analysis pertinent to cloud computing needs to be more ethically grounded, if it seeks to contribute authoritatively to wider ethical debates regarding the ethics of new technologies.

6.2   An Overview of Ethics and Ethical Theories for Analysis of Cloud Computing

Before reviewing existing literature about the ethics of cloud computing, it is important to understand what ethics is, and who the main thinkers that nourished the debate in this branch of philosophy are. Throughout history, many thinkers devoted attention to human actions and asked fundamental questions about their meaning and purpose, reaching very different—and even contrasting—conclusions. For example, Aristotle regarded ethics as a discipline to study the way human beings live and act in order to achieve a life worth living. Aristotle's inquiry was not only theoretical, but also eminently practical: he focused his attention to studying the habitual dispositions that human beings can develop when working towards the realization of their good, which he called virtues (Aristotle 2000). He is considered one of the fathers of Virtue Ethics, one of the most prominent ethical theories. Other thinkers studied questions similar to those of Aristotle, but arrived at different conclusions. The seminal work of Jeremy Bentham and John Stuart Mill, as well as Immanuel Kant, considered to be the fathers of Utilitarianism and Deontological Ethics, respectively, is also very significant. For Bentham, the guiding principle for human actions was the principle of utility, which he first formulated as the achievement of the greatest good for the greatest number of people (Crimmins 2019). Utilitarianism established itself as one of the most pervasive normative ethical theories and considers as "right" that which achieves the satisfaction of the highest number of people possible. Its main focus is on the consequences of human actions, up to the point of considering undesirable actions as "right" if they can in turn achieve desirable outcomes. Despite recognized shortcomings, Utilitarianism inspired many theoretical constructs, particularly in economics and law, thereby influencing economic and societal life. The proponent of the third highlighted approach, Kant, formulated categorical imperatives for every human being to obey, whatever the circumstances and whatever the outcome of the actions following these imperatives. This perspective, more widely known as Deontological Ethics, focused more on rules that need to be respected than on the outcome of actions (Hill 2006). Kant's approach has been highly influential in relation to applied and professional ethics. The challenge pertaining to this approach relates to the identification of both the rules to be followed and who is in charge of setting them.

According to the philosopher Julia Annas (Annas 1993), the main difference between the abovementioned approaches to ethics is that ancient ethical theories (i.e. Aristotelian Virtue Ethics) tend to be centred on the agent, considering the life of the acting person as a narrative unity with a *telos*, a purpose. This agent-centred perspective also characterizes the work of the later Neo-Aristotelian philosophers (such as Anscombe or MacIntyre). On the other hand, more modern ethical theories (i.e. Utilitarianism and Deontological Ethics) are centred on particular actions, and tend to perceive morality as 'punitive or corrective' (Annas 1993, p. 4).

This chronological and thematic overview of the three most prominent approaches to ethics provides a foundation for ethical analysis of cloud computing. However, in reviewing existing contributions which claim to devote their attention to ethical issues related to cloud computing, it appears that these existing contributions: (i) do not explicitly align to a particular school of ethics; (ii) do not offer an explicit or exhaustive account of what ethics is, i.e. do not select a particular definition or approach to ethics; and (iii) do not disclose conceptions of what is "good" or what is "right," which should be the starting point for an analysis which seeks to determine the boundary of what is acceptable (or not) within a professional or business domain.

As a consequence, the structure of the chapter does not reflect the differences between the abovementioned approaches to ethics. The next section, Sect. 6.3, appraises recurring topics in the literature that generally deal with ethical issues related to cloud computing: Section 6.3.1 examines recurring ethical issues linked to the collection, storage and usage of data, Sect. 6.3.2 analyses the perspective of providers of cloud computer services, while Sect. 6.3.3 explores issues related to one of the key stakeholders in the cloud computing industry—the final users of cloud computing services.

6.3   ETHICS AND CLOUD COMPUTING

The previous section introduced ethics and some guiding principles for conducting an ethical analysis according to three prominent ethical approaches. This section reviews existing literature describing ethical concerns related to cloud computing. De Bruin and Floridi (2017) refer to the environment, governments, investors, private and business cloud users, and individuals and corporations interacting with private and business users, as key stakeholders of the cloud computing industry. The literature examined in this section addresses the ethical issues faced by the cloud computing industry when dealing with this complex multi-stakeholder context. The three areas reviewed concern data and the perspective of both providers and users.

6.3.1   *The Ethics of Data in Cloud Computing*

If Aristotle, Bentham or Kant were alive today, they would probably study the manner in which human beings deal with the vast amounts of data currently collected, stored and managed via the cloud. Each would likely promote a different perspective in relation to the best way of dealing with data ethics. Herschel and Miori conducted such an exercise and effectively analysed the ethics pertaining to data according to selected perspectives, including Virtue Ethics, Deontological Ethics and Utilitarianism (Herschel and Miori 2017). They highlight how, from a Kantian perspective, the lack of consent for the collection and use of data would be a clear violation of autonomy, and contrary to Kant's understanding of human beings always as ends in themselves, never as means. They also argue that, from a Utilitarian perspective, the analysis of the ethics of data would be very difficult to perform. They assert that it would be particularly difficult to calculate in a unitary way all the pros and cons of the use of data, and of the ultimate beneficiaries of data usage. In relation to the perspective of Virtue Ethics, this task does not seem to be any easier, given that a Virtue Ethics-based ethical analysis would need to consider how a virtuous person could make the best possible use of data while becoming the best version of themselves.

It would seem that the task of a comprehensive ethical analysis using prominent ethical approaches is particularly difficult, to the point that "data ethics" is now constituting itself as a new branch of applied ethics (Floridi and Taddeo 2016), which is developing its own language and tools, highly correlated with the ethics of algorithms (Mittelstadt et al. 2016). The analysis conducted in this growing—but still very young—field can be applied, *mutatis mutandis*, to the ethics of data in cloud computing. However, the contributions examined in this section make use of a more legalistic framework, strongly influenced by abundant studies in relation to data ownership, and data safety and privacy. As a result, this section offers an overview of these areas related to the ethics of data in cloud computing, which, in the future, will need to incorporate more explicit ethical considerations.

6.3.1.1 Data Ownership

When an individual or a business uses a cloud computing-based platform, it may be asked "who owns the data"? Can this data be considered as private property, or does the mere fact of using a cloud computing platform mean that this data automatically may belong to another party?

The majority of individual and corporate consumers outsource data storage to cloud services, whereby users can use the flexibility and scalability of the cloud without purchasing standalone software or hardware. These are owned and maintained by various service providers whose overall remit is to store and share the data of a multitude of users. However, these service providers do not provide a uniform service. Further, cloud computing services typically traverse national borders, operating in a global context. This global and international cloud computing environment presents difficulties for regulating in such a context. In practice, national laws and regulations may not always align seamlessly into the international domain. As a result, it is becoming apparent that current legal provisions, which are largely pertinent to national jurisdictions, may not appropriately regulate for the cloud (Bartolini et al. 2018). As a consequence, the relevance of ethics is increasingly debated in relation to a potentially essential role regarding the 'outsourced' and international exchange of data in the cloud.

Indeed, the ownership of data and the respect of the right of this particular kind of property is essentially linked with the respect of the fundamental dignity of the human person. The European Data Protection Supervisor (EDPS), an independent institution of the European Union, clearly affirms that human dignity is at the heart of digital ethics: 'the dignity of the human person is not only a fundamental right in itself but also is a foundation for subsequent freedoms and rights, including the rights to privacy and to the protection of personal data' (European Data Protection

Supervisor 2015, p. 12). This authority specifically refers to cloud computing as one of the technologies that needs to carefully address the protection of data stored in cloud based systems, especially in an age when people are requested to upload their data from many and different instances in order to access even services related to basic needs. The EDPS also states that human dignity and ethics can be protected only in so far as the following four pillars are established: current regulation should be future-oriented; accountability of those in charge of checking compliance with internal policies and general regulation should be enhanced by codes of conduct, corporate rules and audits; the computer engineering system should be respectful of human dignity, structurally taking into account issues related to privacy; final users need to be empowered (Sect. 6.3.3 will address this last point more in-depth).

It is intended that these pillars should be applied to cloud computing, especially because issues pertaining to data ownership in the cloud environment may not be clearcut (Al-Khouri 2012; Grimes et al. 2009). Indeed, ownership of data is dependent on the nature of the data owned and where and/or how it was created. Some data is created by the user before uploading to the cloud while other data may be created on the cloud platform (e.g. statistical data). The service provider's terms of services can vary and may grant ownership of such data to the provider or to the public domain i.e. the mere uploading of content to the cloud may erode the user's ownership entitlements to the data (Al-Khouri 2012).

In addition, the arrangement and structuring of data on the cloud may be dominated by the service provider rather than the user and this manipulation of the data, e.g. generating and running algorithms while optimising data or generating statistical analysis, can have implications in relation to ownership. Ultimately, it is difficult to determine who actually owns this optimized data as any data on a cloud platform is likely to have complicated ownership (Cavoukian 2008).

6.3.1.2 Data Security and Privacy

Data security in the context of cloud computing refers to securing data from unauthorised access and is largely a technical issue which providers must implement and maintain (Zissis and Lekkas 2012). Firdhous et al. (2012) discuss the importance of the security issue in specific relation to cloud computing. The complex interconnection of multiple services by a series of different providers for ever increasing numbers of users generates a myriad of issues in relation to the security of users' data. In reality, the strength of the cloud correlates with the security strength, or otherwise, of its weakest actor and a breach or an unauthorised access may effectively impact all users (Ali et al. 2015).

Data security may be regarded as an ethical issue because of the responsibility of those in charge of data collection, storage and usage towards the multi-stakeholder environment involved in the cloud environment. The concept of responsibility has a strong ethical connotation when it needs to go beyond what is prescribed by existing laws, leaving to the willingness of individuals and companies the duty to respect the rights of the owners of the data. Each of the abovementioned ethical theories would agree with the suggestion that data security concurs to the good of data security concurs to the good of society (when data is not related to illegal or unethical issues). This is evidenced more widely in the context of regulation, where various legislative provisions pertaining to data issues in the complex cloud environment have been readily introduced. This highlights that it is relatively more straightforward in some instances to identify a shared legal minimum requirement with regard to data security.

Ethical issues related to data security, when data is not collected, stored and managed by the same entity, are particularly significant. Within the overall cloud context, data security is effectively outsourced to service providers. Security measures adopted are dependent on the delivery models e.g. for SaaS models (i.e. Software-as-a-Service), users depend entirely on service providers to prevent multiple users viewing each other's data, while in PaaS models (i.e. Platform-as-a-Service), providers may assign some security elements to those charged with building applications on top of the platform (Subashini and Kavitha 2011). Reed et al. (2011) highlight six areas of focus in relation to the lifecycle data in the cloud, 'Create', 'Store', 'Use', 'Share', 'Archive' and 'Destroy' and assert that data security measures must be implemented at all stages. The importance of security measures in relation to 'data remanence', the 'residual physical representation of the data after it has been deleted' is also detailed (Kumar et al. 2018, p. 693).

In addition, consumers expect providers to facilitate key data properties: 'integrity', 'confidentiality', and 'availability' (Izang et al. 2017; Kumar et al. 2018; Sun et al. 2014; Tanenbaum and van Steen 2016; Zissis and Lekkas 2012). Integrity of data assumes a confidence that the data has not been manipulated or deleted by unauthorised actors; confidentiality assumes data has not been revealed to unauthorised parties and availability assumes the data is intact and that users can use or recover it as needed. The maintenance of these three properties requires service providers to carefully monitor access and authorisation in respect of all cloud components and may also require linkages to or some supervisory element of third party mechanisms, along with some ongoing verification of the integrity of the data (Bowers et al. 2009; Schiffman et al. 2010). Effective management of all three properties can help to increase overall trust in the system (see below Sect. 6.3.2.1).

The integrity, confidentiality and availability of data is also an essential component of the perception of data security from the point of the users of cloud computing services; this perception is what is characterized as data privacy.

Early concerns regarding privacy issues linked to storing and sharing content online have been somewhat sidelined as the usability and convenience of such systems have proved very attractive in the marketplace (Constantine 2012). However, notwithstanding the wider debate in relation to the moral justification of a right to privacy, the privacy debate in the context of technological innovations is topical (Constantine 2012; Nissenbaum 2009, 2011; Timmermans et al. 2010; van den Hoven 2008). *Privacy* is 'the ability of an individual or group to seclude themselves or information about themselves and thereby reveal themselves selectively' (Sun et al. 2014, p. 6). In many societies, legislation and regulation require compliance such that personally identifiable information is appropriately managed (e.g. European Commission 2018; US Government 1986). In turn, *data privacy* is concerned with the proper handling of data and issues such as consent, notice and compliance with legislation and regulations are dominant (The Centre for Information Policy Leadership 2018). Service providers are thereby challenged with operating services which afford privacy, comply with legal requirements and also balance usability (Stark and Tierney 2014).

In relation to the cloud specifically, Whitworth and de Moor (2003) highlighted conflicts between data privacy and usability at an earlier point. While many of these have subsequently been addressed (Pearson et al. 2009), Stark and Tierney (2014) contend that data in the cloud is 'still too mobile, too "promiscuous" and too often subject to inappropriate use or abuse' (Stark and Tierney 2014, p. 6). They acknowledge that in today's technological society, user autonomy and empowerment must be maximized, but argue that the "safety" of data should not be compromised and highlight mechanisms such as encryption which are widely used to foster privacy within information flows. They also assert that providers must work to ensure data is "live" only when the user is "live" on the network: they stress that the "liveness" of data online, linked to the 'input of the live user, ties decisions about stored online data to an individual' (2014, p. 6). Stark and Tierney (2014) also detail cases where service providers track user information and use this for their own benefit and that of third parties, without the explicit permission or knowledge of users, highlighting that service providers may readily access user data and may also release such data to external agencies.

However, recent governmental policy decisions incorporate focus on such matters e.g. the UK Government's Data Ethics Framework 'sets out clear principles for how data should be used…It will help [organisations] maximise the value of data whilst also setting the highest standards for transparency and accountability when building or buying new data technology' (The Centre for Information Policy Leadership 2018, p. 19). This framework for data ethics is in line with the fact that 'the principle that personal data should be processed only in ways compatible with the specific purpose(s) for which they were collected is essential to respecting individuals' legitimate expectations' (European Data Protection Supervisor 2015, p. 10), thus confirming the idea that regulation should be ethically aware and informed, so that a human-centric use of data would be the only viable possibility.

6.3.2   The Ethics of Providers of Cloud Computing Services

The exploration of ethical and regulatory issues pertinent to data in cloud computing highlights how different stakeholders may have different vested interests in cloud computing and consequently encounter different kinds of ethical issues. Both cloud computing providers and cloud computing users are central to the discussion. Cloud computing providers may be regarded as those who offer cloud-based services and solutions to businesses and/or individuals. They provide virtual hardware, software, infrastructure and other related services. In turn, a cloud user may be regarded as someone who is available to (or who, lacking personal computing resources, necessarily needs to) surrender some control of the data he or she inputs in a system, on the basis that the user can trust the system itself.

For the cloud provider-user relationship to operate effectively, it is critical that cloud providers maintain the trust of users: when a user does not trust the system, he or she is less inclined to expose his or her vulnerability to the system and thereby less inclined to engage with it. The theme of trust is therefore central to cloud computing, and this section begins by examining how ethics can promote the generation of trust, especially through the establishment of codes of ethics—which would need to be respected and enforced beyond the borders of country-based regulation (Sect. 6.3.2.1). The section then considers the perspective of providers regarding sustainability, and the extension of the discussion on "green" computing to cloud computing (Sect. 6.3.2.2). It also explores issues pertaining to the increasing concentration of power and the emergence of oligopolies in the cloud computing industry (Sect. 6.3.2.3).

6.3.2.1   Codes of Ethics for Cloud Computing: Promoting Trust

The accelerated velocity in cloud computing innovation makes it difficult for cloud computing regulators to keep pace with updating standards and guidance and for providers to ensure compliance while maintaining users' trust. Indeed, while it may be debated that 'certification, audits and a new generation of contractual clauses and binding corporate rules can help build a robust trust' (EDPS 2015, p. 10), it may also be argued that 'eliminating bureaucracy in data protection law, by minimising the requirements for unnecessary documentation to maximise room for more responsible initiative by businesses' supported by additional broader guidance would be a more useful future initiative within the sector (EDPS 2015, p. 10).

In the wake of this call for a more responsible implementation of good practices—as opposed to a detailed menu of further regulations—it is interesting to consider the work of Whitehouse et al. (2016), who highlight that further major shifts in the cloud computing landscape are likely in the coming years and advocate heightened emphasis on ethics is necessary with respect to key stakeholders, namely providers and users. This, they suggest, is particularly important if certain behaviours on the part of these stakeholders are desirable into the future. They note that an ethics sufficiently robust to be able to withstand the intensity of the sector and capable of keeping pace with the inevitable developments in the sector is needed but has not yet been established. This, they assert, will help to generate increased trust on the part of users. The International Federation for Information Processing (IFIP) comprises more than 50 different Information and Communication Technology (ICT) global groups. One of their key foci is the societal and ethical challenges of cloud computing (Whitehouse et al. 2016). They assert that enhanced ethics offers many further advantages for the cloud computing sector e.g. improved products and services, improved workplace behaviours, improved overall quality.

One of the potential key constituents for the promotion of a responsible way of building a culture of trust is the introduction of codes of ethics specific to cloud computing (Kouatli 2016). Whitehouse et al. (2016) describe how actors and organisations can make visible their ethical responsibilities and pledge to meet them in a transparent manner within codes. Codes may comprise both codes of conduct, which provide guidance in relation to personal conduct of individuals, and codes of ethics which go beyond the behaviour of individual actors, pertaining to organisations and wider sectoral behaviour. Both involve emphasis on moral guiding principles e.g. 'Contribute to society and well-being'; 'Be honest and trustworthy'; 'Respect the privacy of others'; 'Give proper credit for intellectual property' (Association for Computing Machinery 1978 Code, cited in Whitehouse et al. 2016). Codes thereby are heavily influenced by Deontological Ethics, which focuses on particular imperatives that are typically constant over time. Codes normally also provide more contextual examples to explain how these imperatives might be applied and these examples might be updated periodically to illustrate the application of imperatives as pertinent issues change over time e.g. technological advances. However, codes also focus on balancing different stakeholders' concerns, and, in so doing, incorporate key Utilitarianism components.

Codes have the capacity to support and affirm common standards that go beyond cultural differences of users. This is particularly important to the cloud computing sector, given its global context and the acknowledged difficulties in legislating and regulating in an international environment where different legal, cultural and operational norms may operate in the different jurisdictions and where trust considerations are dependent both on the sector and the country (Fujitsu Research Institute 2010).

Whitehouse et al. (2016) suggest that two-part codes of ethics might be considered, which distinguish between principles and guidance that are internationally acceptable and those that are locally feasible. It is likely that reporting of and monitoring of this guidance may be considered a local issue, given jurisdictional variations. Further, drafting of codes must be general enough to adapt to new and changing conditions without being so general that they cannot be of any practical use. They must also be specific enough to provide direction to relevant actors to the extent of being able to direct them and to also enable stakeholders to hold them to account.

Whitehouse et al. (2016) state that a further 'conscientious, applied examination of key ethical and societal challenges 'will continue to take place in the future both in practice and on practice' (2016, p. 24) to ensure relevant codes can be maintained. However, it is also important to note that criticisms in relation to codes of ethics and their adoption (see e.g. Webley and Werner 2008) can be applied, *mutatis mutandis*, to cloud computing. It is also important to acknowledge that codes of ethics are often perceived as ineffective in changing organizational culture, being necessary but not sufficient measures to exert change in behaviours (Webley and Werner 2008).

6.3.2.2 Green and Sustainable Cloud Computing

Literature on cloud computing has developed notably with reference to "green" computing, which plays to a renewed societal interest in relation to protecting the environment. In some instances, the "green" issue is discussed to the extent that the relatively novel term, "greenwashing", has been linked to this sector (Nanath in Moorthy et al. 2015). This refers to the marketing of cloud computing as more environmentally friendly than other tools, when in reality no substantial effort has been made to render it so. In a more extreme sense, "greenwashing" may even refer to an attempt to make something that is environmentally damaging appear to be environmentally friendly. This section reports different perspectives on cloud computing providers and their attention (or lack thereof) to environmentally sensitive issues.

Scott and Watson (2012) assert that new technologies such as cloud computing 'provide unprecedented opportunities to improve the efficiency of business operations and represent a realistic opportunity to reduce energy costs and combat global warming' (Scott and Watson 2012, p. 1). Di Salvo et al. (2017) go so far as to affirm that cloud computing is not only a green technology, but it is the "greenest." They claim that the centralised aspects of cloud computing effectively facilitate the operation of a virtual machine system and storage of data in a manner that consumes less than half the energy and global resources required by traditional decentralised systems. Lin (2012) also examines "green" cloud computing and highlights that a well-considered scheduling strategy of the various information flows, combined with careful management of resource allocation, may further contribute to a more sustainable and green technology.

Buyya and Gill (2018) provide a more balanced discussion, arguing that while cloud computing reduces some discrete environmental footprint, the wider data center industry may have a disproportionate negative environmental impact. In practice, hyperscale cloud data centers consume a disproportionate amount of energy—currently almost on a par with sectors such as the aviation sector—and are reaching technical limits. This negative impact of high carbon usage on global warming has been referred to as the "dark side" to cloud computing.

Bachour and Chasteen (2010, p. 1) refer to the 'triple bottom line' concept and the monitoring of 'economic viability', 'social responsibility' and 'environmental impact,' which they acknowledge may be difficult to define and measure. Scott and Watson (2012) urge caution in relation to monitoring and measuring green issues connected to cloud computing and other technologies. They assert that potential models for identifying and measuring such "green" components must consider a wide range of impacts, including direct financial impacts as well as broader societal impacts. They propose a framework which captures key stakeholders—societal, organizational, customer and supply-chain—and core value dimensions—economic, environmental, ethical and competitive position. Buyya and Gill (2018) advocate an alternative "green" model for holistic management of resources to ensure more energy-efficient and sustainable cloud operations in a bid to further decrease carbon footprints of cloud datacentres. Similarly, Di Salvo et al. (2017) note that the extent of "green" labelling has to date been based solely on energy consumption reduction and assert a more appropriate assessment of sustainable practices should also consider wider aspects such as eco-energy efficiency performance.

The duality of the cloud and its impact on the environment is thereby acknowledged: cloud computing has much potential to impact positively on discrete environmental matters; however, less positive 'green' computing considerations also need to be addressed. While most climate change activists are currently focused on limiting emissions from other sectors, such as the auto, aviation and energy sectors, the cloud computing industry may be on track to generate more carbon emissions than all of these sectors combined. It is therefore important that this "green" issue be even a more central component of the next generation of cloud (Garg et al. 2015; Lin 2012; Mohamed and Pillutla 2014; Murugesan 2008).

6.3.2.3 Concentration of Power and Oligopolies in the Cloud Computing Industry

As previously highlighted, the cloud overarches national and international borders and boundaries. Indeed, it may be observed that cloud computing has accelerated the pace of a truly global world (Mohamed and Pillutla 2014). While the cloud is offered to users all over the world, cloud computing as an industry is very much concentrated in the hands of a relatively small number of providers, akin to an oligopolistic market. These providers sustain the cost of the cloud infrastructure and decide on the cloud services provided. An issue commonly associated with oligopolies is the control that the small number of companies providing a specific good or service have on setting prices. This concern is also prevalent with regard to cloud computing: Feng et al. 2014 conduct an analysis of cloud providers' pricing strategy, addressing the cloud computer economic arena as an oligopoly market. Timmermans et al. (2010) assert that a significant amount of cloud computing innovation originates in the US. Synergy Research Group (Alley 2019) tracks such activity and reports that 38% of worldwide hyperdata centres are located in the US. It also reports that Amazon, Microsoft, Google, Alibaba and Oracle are the most active companies in this arena. These are effectively stockpiling information about consumers, presumably seeking to gain some competitive advantage in the process.

In these aforementioned contributions, there is no explicit mention of ethical issues associated with the concentration of power in a key industry such as cloud computing. Usually, ethical issues related to concentration of power regard the (un)fair treatment of the weak party in the economic agreement (Crane and Matten 2016) i.e. powerful oligopolies can establish terms and conditions that violate the rights of final users, who are effectively forced to accept them if they want to benefit from the service. Whatever the selected approach to ethics might be, a concentration of power is usually not in favour of the satisfaction of the greatest number of people (Utilitarianism), nor of the respect of established rules (Deontological Ethics).

Moreover, concentration of power which generates oligopolies is also responsible for spreading particular cultural influences. Timmermans et al. (2010) reflect on such trends within cloud computing and ponder if Western values may thereby dominate pertinent future applications, frameworks and regulations, and lead to 'increasing cultural homogenization' (2010, p. 5). This argument is discussed by Ess (2001) who posits that today's multicultural domains 'reflect an extensive diversity, if not cacophony, of cultural identities, traditions, voices, views, and practices' (2001, p. 196) and asserts that it is critically important that cloud systems deal with such diversity in an ethical manner. However, if current trends continue, it is likely that Western perspectives will dominate into the future.

6.3.3   The Ethics of Cloud Computing Users

The ethical issues pertaining to users of cloud computing services predominantly relate to the way they use the services. Cloud users are the main stakeholders of cloud provider entities, being at the same time the most vulnerable element of the complex chain of interactions that cloud computing services entail. With reference to the ultimate technical capabilities and limitations imposed by the different cloud systems, users decide how to interact relative to the context in which they find themselves. One means of understanding the perspective of users of cloud computing services is to engage with the term "prosumers." The expression was coined in 1980 by Alvin Toffler (Toffler 1980), and refers to consumers who, simultaneously, are also producers. The European Data Protection Supervisor discusses this term in relation to the end users of new technologies, including cloud computing. The associated reference to prosumers relates to one of the pillars for a new digital ethics: the empowerment of individuals using technologies. Consumers are no longer passive receptors of the available good or service; rather, they co-create the product present in digital platforms. As a result, the empowerment of twenty-first century prosumers is meant to increase their technological and ethical awareness. This perspective of empowerment is a top-down approach, where regulators are attempting to provide tools to strengthen the role and responsibility of individual users of digital platforms.

An alternative perspective on users' awareness of their digital presence is to consider how individuals can empower themselves, without the support of a regulatory authority. Cloud users themselves can learn how to engage with the cloud environment by developing intellectual and practical habits that may be examined through the lens of the epistemic virtues. De Bruin and Floridi (2017) examine epistemic virtues, also referred to as informational or knowledge virtues, in light of the specific circumstances that cloud computing users encounter. They highlight that users must be able to make informed decisions in relation to their engagement with cloud services. To enable such informed decisions, there must be clarity of communication between the two parties i.e. between providers and users. De Bruin and Floridi (2017) note current barriers to communication within the industry whereby terms of usage are written in lengthy and complex legal jargon which many people do not readily understand.

De Bruin and Floridi (2017), building on the work of Kawall (2002) and De Bruin (2015), ultimately characterize the virtue of "interlucency" as the other-regarding epistemic virtue that helps establish a common knowledge through providing the relevant information in a language that can be understood and shared by the interested users. More than the virtue of a single cloud user, the epistemic virtue of interlucency, according to De Bruin and Floridi, must be applied to the cloud computing industry as a whole. To ground their contribution, the authors make explicit reference to the concept of "comprehensibility" as formulated by Habermas (quoting Habermas 1981).

De Bruin and Floridi (2017) further discuss users' attitudes. They suggest that users should be intellectually impartial, sober and courageous. Intellectual impartiality applies to users insofar as they are able to overcome personal prejudices towards the cloud computing industry and the possible risks connected to the circulation of their data. They may attempt to do so by listening to authoritative voices in the field of technology and by seeking information from independent experts. Intellectual sobriety helps users to 'resist the overly enthusiastic adoption of beliefs about either the pros and the cons of cloud computing' (2017, p. 29). Intellectual courage concerns the continuous active attitude to overcome ignorance and lack of information even when faced with obstacles. The authors assert that users must continue to ask questions when something is not clear, for example the terms and conditions of a specific service.

It is interesting to note how De Bruin and Floridi (2017) articulate in favour of "proscriptive pressure" on the side of users more than on the side of the providers of cloud services: in so doing, they are actually interpreting one of the main ethical issues associated with the relationship between technology and the use of technology. Indeed, the authors place emphasis on the freedom of the agent and on his/her responsibility to make the best use of technology without limiting the possibility of technology developers to advance their discoveries, highlighting that the ethical boundaries of technological advances should be determined in relation to the way they are used, not by what they can potentially do. For example, a terrorist organization can benefit from services available in the cloud, however the objective for which it uses this service is illegal and contrary to the good of society.

6.4   Conclusions

The chapter offers an overview of ethics and prominent ethical theories, which can help to chart existing literature in relation to ethics and cloud computing. One of the key findings of this analysis is that the contributions aimed at addressing ethical issues in cloud computing do not explicitly refer to a specific conception of ethics, nor explicitly declare a reference point for their analysis from an ethical perspective. It was therefore not possible to structure the chapter in line with the different schools of ethics, and required analyses to focus on recurring topics within selected cloud computing literature. Given the complex multi-stakeholder cloud computing environment, ethical issues particular to cloud computing were distinguished in relation to three categories: data, providers and users.

With reference to data, issues related to ownership, security and privacy were discussed. The cloud computing literature in this context is particularly technical and oriented to address the legal framework of pertinent data issues. Further research in relation to data ethics in cloud computing may be inspired by the seminal work conducted in the parallel field of data ethics and algorithm ethics. In addition, the methodology of social ethics could be useful to conduct research on the way the society of the future will store and manage data via cloud computing technology. This research could explore the different kinds of society that could result depending on the quality and quantity of individual and collective data available and how human beings could live in these kinds of societies. Another interesting direction that the research about data ethics in cloud computing might take is suggested by Koehn (2019). In the context of the new technologies of the twenty-first century, she argues that it is worth reflecting on Plato's Socratic notion of responsibility: this notion is essentially dialogic and goes beyond the concept of responsibility typical of the Anglo-American culture, which perceives responsibility as part of a role-based duty. This personalization of responsibility can help those who manage data on others' behalf, so that they do not limit themselves to what is prescribed by their role, but are able to see the relational component of their role towards the people on the cloud platform, considering them as people and not as a set of data.

With reference to the ethical issues related to cloud computing service providers, codes of ethics, green and sustainable cloud computing, and concentration of power were examined. Future research in sustainability and oligopolies in cloud computing may consider studies such as those conducted in mainstream literature regarding these topics, while ethics and cloud computing literature might explore further possibilities for writing and implementing codes of ethics, and may seek to examine the experiences (and failures) of other sectors in this regard.

As regards ethical issues pertaining to cloud users, the perspectives of individual empowerment and epistemic virtues were explored. Very few authors have dedicated attention to these important areas. A possible pathway for future research may be an application of ancient Aristotelian ethics to the new technological environment: while the context may have profoundly changed, the fundamental attitudes and questions of human beings still remain and the ethics and cloud computing literature could benefit from the wisdom of one of the first ethical theories formulated.

It may be argued that the pace of academic publications, while slower than cloud computing innovation, is still relatively rapid. It must be acknowledged that the majority of contributions examined in this chapter report specific perspectives on focussed topics in the cloud computing domain, without delving further into the theoretical foundations of their ethical analysis. It is anticipated that the structure and articulation of this chapter will aid the reader to understand how a multi-stakeholder ethical analysis is needed in order to appropriately examine the multi-stakeholder context within the cloud computing domain. The aforementioned categories selected for this review are interrelated and overlapping, and highlight factual complexities. The cloud providers and users must be mindful that they shape the morality of the cloud computing context through their interactions with the other stakeholders and with the platform itself, and they must not abdicate their responsibilities to the possibilities offered by the technology. Providers must be accountable, not only to the final users of their platforms, but also to the environment, and to the cultural context within which the platforms operate. It must also be remembered that regulation, although continuously evolving, is unlikely to advance at the pace of cloud computing innovation and this review clearly illustrates that ethics is an integral component to orient cloud computing towards the good of society into the future.

REFERENCES

Ali, M., Khan, S. U., & Vasilakos, A. V. (2015). Security in Cloud Computing: Opportunities and Challenges. *Information Sciences, 305*, 357–383. https://doi.org/10.1016/j.ins.2015.01.025.

Al-Khouri, A. M. (2012). Data Ownership: Who Owns "My Data"? *International Journal of Management Information Technology, 2*(1), 1–8. https://doi.org/10.24297/ijmit.v2i1.1406.

Alley, A. (2019, October 18). *Synergy: There Are More Than 500 Hyperscale Data Centers in the World.* Retrieved from https://www.datacenterdynamics.com/news/synergy-there-are-more-500-hyperscale-data-centers-world/

Annas, J. (1993). *The Morality of Happiness.* Oxford University Press. https://doi.org/10.1093/0195096525.001.0001.

Aristotle. (2000). *Nicomachean Ethics.* Cambridge, UK and New York: Cambridge University Press.

Bachour, N., & Chasteen, L. (2010). *Optimizing the Value of Green IT Projects within Organizations.* 2010 IEEE Green Technologies Conference, pp. 1–10. https://doi.org/10.1109/GREEN.2010.5453804

Bartolini, C., Santos, C., & Ullrich, C. (2018). Property and the Cloud. *Computer Law & Security Review, 34*(2), 358–390. https://doi.org/10.1016/j.clsr.2017.10.005.

Bowers, K. D., Juels, A., & Oprea, A. (2009). *HAIL: A High-Availability and Integrity Layer for Cloud Storage.* Proceedings of the 16th ACM Conference on Computer and Communications Security, pp. 187–198. https://doi.org/10.1145/1653662.1653686

Buyya, R., & Gill, S. S. (2018). Sustainable Cloud Computing: Foundations and Future Directions. *Business Technology & Digital Transformation Strategies, 21*(6), 1–9.

Cavoukian, A. (2008). Privacy in the Clouds. *Identity in the Information Society, 1*(1), 89–108. https://doi.org/10.1007/s12394-008-0005-z.

Constantine, D. (2012). Cloud Computing: The Next Great Technological Innovation, the Death of Online Privacy, or Both. *Georgia State Law Review, 499,* 499–528.

Crane, A., & Matten, D. (2016). *Business Ethics: Managing Corporate Citizenship and Sustainability in the Age of Globalization* (4th ed.). Oxford: Oxford University Press.

Crimmins, J. E. (2019). Jeremy Bentham. In E. N. Zalta (Ed.), *Stanford Encyclopedia of Philosophy.* Retrieved from https://plato.stanford.edu/archives/sum2019/entries/bentham/

De Bruin, B. (2015). *Ethics and The Global Financial Crisis: Why Incompetence Is Worse Than Greed.* Cambridge: Cambridge University Press.

de Bruin, B., & Floridi, L. (2017). The Ethics of Cloud Computing. *Science and Engineering Ethics, 23*(1), 21–39. https://doi.org/10.1007/s11948-016-9759-0.

Di Salvo, A. L. A., Agostinho, F., Almeida, C. M. V. B., & Giannetti, B. F. (2017). Can Cloud Computing Be Labeled as "Green"? Insights Under an Environmental Accounting Perspective. *Renewable and Sustainable Energy Reviews, 69,* 514–526. https://doi.org/10.1016/j.rser.2016.11.153.

Ess, C. (2001). Culture and Global Networks: Hope for a Global Ethics? In J. van den Hoven & J. Weckert (Eds.), *Information Technology and Moral Philosophy* (pp. 195–225). Cambridge: Cambridge University Press. https://doi.org/10.1017/CBO9780511498725.012.

European Commission. (2018). *General Data Protection Regulation*. Retrieved from https://eur-lex.europa.eu/legal-content/EN/TXT/?uri=COM:2018:43:FIN

European Data Protection Supervisor. (2015, September 11). *Towards a New Digital Ethics*. Retrieved from https://edps.europa.eu/sites/edp/files/publication/15-09-11_data_ethics_en.pdf

Feng, Y., Li, B., & Li, B. (2014). Price Competition in an Oligopoly Market with Multiple IaaS Cloud Providers. *IEEE Transactions on Computers, 63*(1), 59–73. https://doi.org/10.1109/TC.2013.153.

Firdhous, M., Ghazali, O., & Hassan, S. (2012). Trust Management in Cloud Computing: A Critical Review. *International Journal on Advances in ICT for Emerging Regions (ICTer), 4*(2), 24. https://doi.org/10.4038/icter.v4i2.4674.

Floridi, L., & Taddeo, M. (2016). What is Data Ethics? *Philosophical Transactions of the Royal Society A: Mathematical, Physical and Engineering Sciences, 374*(2083), 20160360. https://doi.org/10.1098/rsta.2016.0360.

Fujitsu Research Institute. (2010). *Personal Data in the Cloud: A Global Survey of Consumer Attitudes*. Retrieved from https://www.fujitsu.com/uk/Images/fujitsu_personal-data-in-the-cloud.pdf

Garg, S., Dwivedi, R. K., & Chauhan, H. (2015). *Efficient Utilization of Virtual Machines in Cloud Computing Using Synchronized Throttled Load Balancing.* 2015 1st International Conference on Next Generation Computing Technologies (NGCT), pp. 77–80. https://doi.org/10.1109/NGCT.2015.7375086

Grimes, J. M., Jaeger, P. T., & Lin, J. (2009). *Weathering the Storm: The Policy Implications of Cloud Computing*. Illinois Digital Environment for Access to Learning and Scholarship Conference.

Habermas, J. (1981). *Theorie des kommunikativen Handelns*. Frankfurt am Main: Suhrkamp.

Herschel, R., & Miori, V. M. (2017). Ethics & Big Data. *Technology in Society, 49*, 31–36. https://doi.org/10.1016/j.techsoc.2017.03.003.

Hill, T. E., Jr. (2006). Kantian Normative Ethics. In D. Copp (Ed.), *The Oxford Handbook of Ethical Theory* (pp. 480–514). New York: Oxford University Press.

Izang, A. A., Adebayo, A. O., Okoro, O. J., & Taiwo, O. O. (2017). Security and Ethical Issues to Cloud Database. *The Journal of Computer Science and Its Applications, 24*(2), 65–75.

Kawall, J. (2002). Other–Regarding Epistemic Virtues. *Ratio, 15*(3), 257–275. https://doi.org/10.1111/1467-9329.00190.

Koehn, D. (2019). *Toward a New (Old) Theory of Responsibility: Moving beyond Accountability*. https://doi.org/10.1007/978-3-030-16737-0

Kouatli, I. (2016). Managing Cloud Computing Environment: Gaining Customer Trust with Security and Ethical Management. *Procedia Computer Science, 91*, 412–421. https://doi.org/10.1016/j.procs.2016.07.110.

Kumar, P. R., Raj, P. H., & Jelciana, P. (2018). Exploring Data Security Issues and Solutions in Cloud Computing. *Procedia Computer Science, 125*, 691–697. https://doi.org/10.1016/j.procs.2017.12.089.

Lin, C. (2012). A Novel Green Cloud Computing Framework for Improving System Efficiency. *Physics Procedia, 24*, 2326–2333. https://doi.org/10.1016/j.phpro.2012.02.345.

Mittelstadt, B. D., Allo, P., Taddeo, M., Wachter, S., & Floridi, L. (2016). The Ethics of Algorithms: Mapping the Debate. *Big Data & Society, 3*(2), 205395171667967. https://doi.org/10.1177/2053951716679679.

Mohamed, M., & Pillutla, S. (2014). Cloud Computing: A Collaborative Green Platform for the Knowledge Society. *VINE, 44*(3), 357–374. https://doi.org/10.1108/VINE-07-2013-0038.

Moorthy, J., Lahiri, R., Biswas, N., Sanyal, D., Ranjan, J., Nanath, K., & Ghosh, P. (2015). Big Data: Prospects and Challenges. *Vikalpa: The Journal for Decision Makers, 40*(1), 74–96. https://doi.org/10.1177/0256090915575450.

Murugesan, S. (2008). Harnessing Green IT: Principles and Practices. *IT Professional, 10*(1), 24–33. https://doi.org/10.1109/MITP.2008.10.

Nissenbaum, H. (2009). *Privacy in Context: Technology, Policy, and the Integrity of Social Life*. Stanford, CA: Stanford Law Books.

Nissenbaum, H. (2011). A Contextual Approach to Privacy Online. *Daedalus, 140*(4), 32–48.

Pearson, S., Shen, Y., & Mowbray, M. (2009). A Privacy Manager for Cloud Computing. In M. G. Jaatun, G. Zhao, & C. Rong (Eds.), *Cloud Computing* (Vol. 5931, pp. 90–106). Heidelberg: Springer. https://doi.org/10.1007/978-3-642-10665-1_9.

Reed, A., Rezek, C., & Simmonds, P. (Eds.). (2011). *Security Guidance for Critical Areas of Focus in Cloud Computing V3. 0*. Retrieved from https://downloads.cloudsecurityalliance.org/initiatives/guidance/csaguide.v3.0.pdf

Schiffman, J., Moyer, T., Vijayakumar, H., Jaeger, T., & McDaniel, P. (2010). *Seeding Clouds with Trust Anchors*. Proceedings of the 2010 ACM Workshop on Cloud Computing Security Workshop—CCSW'10, p. 43. https://doi.org/10.1145/1866835.1866843

Scott, M., & Watson, R. (2012). *The Value of Green IT: A Theoretical Framework and Exploratory Assessment of Cloud Computing*. BLED 2012 Proceedings, p. 30.

Stark, L., & Tierney, M. (2014). Lockbox: Mobility, Privacy and Values in Cloud Storage. *Ethics and Information Technology, 16*(1), 1–13. https://doi.org/10.1007/s10676-013-9328-z.

Subashini, S., & Kavitha, V. (2011). A Survey on Security Issues in Service Delivery Models of Cloud Computing. *Journal of Network and Computer Applications, 34*(1), 1–11. https://doi.org/10.1016/j.jnca.2010.07.006.

Sun, Y., Zhang, J., Xiong, Y., & Zhu, G. (2014). Data Security and Privacy in Cloud Computing. *International Journal of Distributed Sensor Networks, 10*(7), 190903. https://doi.org/10.1155/2014/190903.

Tanenbaum, A. S., & van Steen, M. (2016). *Distributed Systems: Principles and Paradigms* (2nd ed., adjusted for digital publishing). Leiden: Maarten van Steen.

The Centre for Information Policy Leadership. (2018). *The Case for Accountability: How it Enables Effective Data Protection and Trust in the Digital Society.* Retrieved from https://www.informationpolicycentre.com/uploads/5/7/1/0/57104281/cipl_accountability_paper_1_-_the_case_for_accountability_-_how_it_enables_effective_data_protection_and_trust_in_the_digital_society.pdf

Timmermans, J., Stahl, B. C., Ikonen, V., & Bozdag, E. (2010). *The Ethics of Cloud Computing: A Conceptual Review.* 2010 IEEE Second International Conference on Cloud Computing Technology and Science, pp. 614–620. https://doi.org/10.1109/CloudCom.2010.59

Toffler, A. (1980). *The Third Wave* (1st ed.). New York: Morrow.

US Government. (1986). *Electronic Communications Privacy Act.* Retrieved from https://www.justice.gov/jmd/electronic-communications-privacy-act-1986-pl-99-508

van den Hoven, J. (2008). Moral Methodology and Information Technology. In K. E. Himma & H. T. Tavani (Eds.), *The Handbook of Information and Computer Ethics* (pp. 49–67). Hoboken, NJ: Wiley. https://doi.org/10.1002/9780470281819.ch3.

Webley, S., & Werner, A. (2008). Corporate Codes of Ethics: Necessary But Not Sufficient. *Business Ethics: A European Review, 17*(4), 405–415. https://doi.org/10.1111/j.1467-8608.2008.00543.x.

Whitehouse, D., Duquenoy, P., Kimppa, K. K., Burmeister, O. K., Gotterbarn, D., Kreps, D., & Patrignani, N. (2016). Twenty-Five Years of ICT and Society: Codes of Ethics and Cloud Computing. *ACM SIGCAS Computers and Society, 45*(3), 18–24. https://doi.org/10.1145/2874239.2874242.

Whitworth, B., & De Moor, A. (2003). Legitimate by Design: Towards Trusted Socio-Technical Systems. *Behaviour & Information Technology, 22*(1), 31–51. https://doi.org/10.1080/01449290301783.

Zissis, D., & Lekkas, D. (2012). Addressing Cloud Computing Security Issues. *Future Generation Computer Systems, 28*(3), 583–592. https://doi.org/10.1016/j.future.2010.12.006.

CHAPTER 7

Trustworthy Cloud Computing

Olasunkanmi Matthew Alofe and Kaniz Fatema

Abstract Trustworthy cloud computing has been a central tenet of the European Union cloud strategy for nearly a decade. This chapter discusses the origins of trustworthy computing and specifically how the goals of trustworthy computing—security and privacy, reliability, and business integrity—are represented in computer science research. We call for further inter- and multi-disciplinary research on trustworthy cloud computing that reflect a more holistic view of trust.

Keywords Trustworthy computing • Cloud computing • Trust • Reliability • Security • Business integrity

O. M. Alofe (✉)
University of Derby, Derby, UK
e-mail: o.alofe1@unimail.derby.ac.uk

K. Fatema
Department of Computer Science, Aston University, Birmingham, UK

© The Author(s) 2021
T. Lynn et al. (eds.), *Data Privacy and Trust in Cloud Computing*,
Palgrave Studies in Digital Business & Enabling Technologies,
https://doi.org/10.1007/978-3-030-54660-1_7

129

7.1   Introduction

In 2002, Bill Gates, in an email to Microsoft employees, presaged a future where computing would be "an integral and indispensable part of almost everything we do" (Gates 2002). Subsequently, Microsoft published a white paper defining what would ultimately become a seminal white paper for trustworthy computing. Recognising that trust is a complex concept, Mundie et al. (2002) explored trustworthy computing from three perspectives—the user's perspective (goals), the mechanisms employed by industry to meet the goals (means), and the way in which an organisation conducts its operations to deliver the components (execution). The key definitions of goals, means and execution are summarised in Table 7.1 below. While in 2002, cloud computing was not the dominant computing paradigm it is today, these perspectives reflect the dominant themes in computer science research on trust in cloud computing. Indeed they are reflective of the wider scholarly debate discussed throughout this book.

Improving the confidence and perception of trustworthiness is critical for the adoption of cloud computing, and has been a central tenet of the European Union cloud strategy for nearly a decade (European Commission 2020). The remainder of this chapter provides a brief overview of computer science research based on the goals of trustworthy computing identified above, namely security and privacy, reliability, and business integrity.

7.2   Security and Privacy

According to the National Information Systems Security Glossary, information security is the protection of information systems against unauthorised access to and modification of information and data in various forms such as data at rest, and in transit (Hayden 2000). Information security applies to the safeguarding of data in its various states and storage locations, as well as offering protection against attacks such as denial-of-service (DoS), which might adversely impact the confidentiality, integrity, and availability of information to authorised users. As discussed in Chap. 1, integrity is a key element in trust, and in the context of cloud computing, the maintenance of confidentiality and continuity of service availability are key signals of competence. As such, from a computer science perspective, designing attack-resilient systems is critical to building and maintaining trust. Different frameworks and models have been proposed and designed for the establishment of trust within cloud computing that offer system

Table 7.1 Definition of goals, means, and execution in trustworthy computing

Goals	The basis for a customer's decision to trust a system
Security & privacy	The expectation of attack-resilient systems and that the confidentiality, integrity, and availability of the system and its data are protected.
	The customer can control data about themselves, and those using such data adhere to fair information principles
Reliability	The customer can depend on the product to fulfil its functions when required to do so.
Business integrity	The vendor of a product behaves in a responsive and responsible manner.
Means	**The business and engineering considerations that enable a system supplier to deliver on the Goals**
Secure by design, secure by default, secure in deployment	A process is in place to protect the confidentiality, integrity, and availability of data and systems at every phase of the software development process.
Fair information principles	The collection and sharing of end-user data requires the consent of the end user, and privacy is respected, and data is only used in line with Fair Information Practices.
Availability	The system is available for use as required.
Manageability	The system is easy to install and manage, relative to its size and complexity.
Accuracy	The system performs its functions correctly. Data is protected from corruption and loss.
Usability	The software is easy to use and suitable to the user's needs.
Responsiveness	The vendor accepts responsibility for problems, and takes action to correct them. Support is available to customers as needed throughout their engagement with vendor.
Transparency	The vendor is open in its dealings with customers. Its motives are clear, it keeps its word, and customers know where they stand in a transaction or interaction with the vendor.
Execution	**The way an organisation conducts its operations to deliver the components required for trustworthy computing**
Intents	• Company policies, directives, benchmarks, and guidelines • Contracts and undertakings with customers, including Service Level Agreements (SLAs) • Corporate, industry and regulatory standards Government legislation, policies, and regulations

(continued)

Table 7.1 (continued)

Implementation	• Risk analysis • Development practices, including architecture, coding, documentation, and testing • Training and education • Terms of business • Marketing and sales practices • Operations practices, including deployment, maintenance, sales & support, and risk management • Enforcement of intents and dispute resolution
Evidence	• Self-assessment • Accreditation by third parties • External audit

Adapted from Mundie et al. (2002)

security and data privacy for cloud service providers and their customers. Five common approaches for protecting cloud systems and data in extant literature include multi-cloud storage, homomorphic encryption schemes, secure sharing systems, deployment of intermediary components, as well as more traditional security and privacy methods.

Multi-cloud storage strategies seek to reduce security and availability risks by diversifying this risk through the use of multiple cloud storage service providers (Bucur et al. 2018). For example, Alqahtani and Kouadri-Mostefaou (2014) propose a framework that ensures the security of mobile cloud computing by deploying distributed multi-cloud storage, data encryption, and data compression techniques. The framework operates by dividing the data into different segments at the user end based on the preference selected by the user before the encryption and compression of the segments. The compressed segments are stored on distributed multi-cloud storage service providers. Similarly, Abdalla and Pathan (2014) presented a framework using a data protection manager (DPM) deployed for the transmission of data to the cloud service provider. The DPM both fragments and merges the data in the proposed framework. First, it breaks the data into fragments and transmits them to the multi-cloud for storage. When a user requests the data, the DPM merges the data. The service provider maps the information of fragmented and merged data to the individual users and the multi-cloud technique applied protects data on other segments if one segment is compromised. While multi-cloud storage in theory has many advantageous attributes, in practice, it has significant limitations, not least the lack of standards-based interoperable clouds and APIs, the possible amplification of the attack surface to multiple clouds, and the management and measurement of multiple service level agreements across multiple clouds (Bucur et al. 2018).

There is a long history of encryption as a means of securing systems. For example, many messaging systems use encryption to protect the content of messages through the use of shared public or private keys. These legacy systems have a number of limitations including data control and the management of keys (Acar et al. 2018). Homomorphic encryption schemes overcome these limitations by allowing a cloud service provider to perform certain computable functions on the encrypted data while preserving the features of the function and format of the encrypted data (Acar et al. 2018). Louk and Lim (2015) proposed a homomorphic data security encryption scheme that converted data into ciphertext and manipulated the ciphertext just like the original text without compromising the encryption. There are a variety of different homographic encryption types, for example multiplicative, additive and fully homomorphic, all of which have been applied to secure communication and storage in the cloud (Tebaa and Hajji 2014). There are significant performance limitations with fully homomorphic encryption schemes thus requiring optimisation at the architectural, algorithmic, and hardware resource levels (Moore et al. 2014).

The ubiquity of smartphones, and their dependence on cloud computing, present significant challenges for securing data at the edge, in the cloud, and in between. Smartphones, and indeed other Internet of Things end points, are typically resource constrained due to their form and bandwidth. As such, security methods need to be relatively lightweight. Wang et al. (2014) propose a secure sharing scheme that envisages users uploading multiple data pieces to different clouds, and using a watermarking algorithm for authentication of mobile users and cloud services. A key feature of this solution is the both the security and the reduced load on the network. Khan et al. (2014) propose a BSS (block-based sharing scheme) cryptographic method that divides data logically into multiple blocks, encrypting and decrypting the blocks, and reconstructing the data into their original form. Secure Data Sharing in Clouds (SeDaSC) is another approach to secure sharing comprising three entities—the user, a cryptographic server (CS) and the cloud (Ali et al. 2015). The CS is responsible for encryption, decryption, key management, and access control. Yu et al. (2015) proposed a public auditing protocol that ensures the integrity of data stored in the cloud and shared data among users by using the asymmetric group key agreement scheme and proxy re-signature. The asymmetric group key agreement scheme allows the group to share both public and private keys and create a tag attached to files. The proxy re-signature updates the tags when there are changes in the group members. User identity information is preserved by anonymising the auditor and group members. In this way, data control is improved, in instances such as when employees leave an organisation.

Similar to the auditing scheme proposed by Yu et al. (2015), a number of works have proposed auditing schemes where, in effect, an independent third party serves as the verifier of data integrity. For example, Sookhak et al (2014) proposed a remote data auditing method for verifying the integrity of data stored in cloud; algebraic signatures are used to allow the auditor to check the possession of user data in cloud. Similarly, Yu et al. (2016) propose key-updating and authenticator-evolving mechanism with zero-knowledge privacy of the stored files for secure cloud data auditing, which incorporates zero-knowledge proof systems, proxy re-signatures and homomorphic linear authenticators. Yang et al. (2015) proposed an extended proxy-assisted approach that utilises an attribute-based encryption method to ensure scalable data sharing within the cloud. Tian et al. (2015) proposed a dynamic hash table (DHT) public auditing scheme. The DHT is a two-dimensional data structure used by the auditor to record data property information for rapid and dynamic auditing. Public key-based homomorphic authentication and random masking created by the auditor are used for the preservation of privacy.

While each of the approaches above represent novel means to securing data, the practical reality is that most cloud service providers rely on traditional security and privacy methodologies. A wide range of approaches have been proposed for securing cloud services including securing infrastructure using extant multi-component methods. For example, Liu et al. (2015) propose a secure infrastructure based on Advanced Encryption Standard (AES), Searchable Symmetric Encryption (SSE), Ciphertext-Policy Attribute-Based Encryption (CPABE) and Digital Signature (DS). Mollah et al. (2017) propose a scheme that utilises a combination of secret key encryption, public key encryption, searchable secret key encryption and digital signatures for a data searching and sharing scheme. The STOVE model proposed by Tan et al. (2014) secures data in the cloud by restricting the operational ability of applications. The model restricts untrusted applications and isolates the application using formal verification methods to verify the isolated code; application execution is performed in isolation and under strict observation. The novelty of these methods, and many others, is in the combination of multiple approaches. However, the challenge for industry and researchers alike is identifying the most feasible candidates for a given use case.

7.3   RELIABILITY

It is essential that services and data in the cloud are available to users at all times. As discussed in Chap. 2, availability is defined in the service level agreements between cloud service providers and their customers. The most commonly used definition of reliability in engineering applications according to Dummer et al. (1997, p. 79) is "the characteristic of an item expressed by the probability that it will perform a required function under stated conditions for a stated period of time." In general terms, service reliability can be represented as:

$$\text{Service Reliability} = \frac{\left(\text{Successful Responses}\right)}{\text{Total Requests}} \times 100\%.$$

While such a calculation may indicate service reliability, in hyperscale multi-tenant clouds the overall cloud may be reliable but specific services may be unreliable. Due to the scale of the clouds, one particular service failure or underperforming component may not impact an overall reliability score, while at the same time result in catastrophic failure. Huang et al. (2017) suggest that major cloud failures often result from subtle underlying faults in systems, so-called 'gray failures', that may be difficult to observe or even detect. They are characterised by this differential observability (Huang et al. 2017).

When ascertaining that a system will perform a specific function within a given cloud service environment, Adams et al. (2014) suggest the following key considerations:

- Service availability must be maximised to ensure users can access the service and perform their required task to completion without interference;

- The impact of system failure should be minimised for individual users, the overall number of users affected, and the downtime associated for the failure;
- Service performance and capacity should be maximised to reduce the impact of reduced performance even if no failure is detected; and,
- Business continuity should be maximised by responding to failures when they occur, protecting the integrity of data, and recovering as soon as possible.

Reliability and high availability are closely related and regarded as significant challenges in cloud computing. Obviously, cloud service providers and scholars invest a significant amount of effort in to the design of fault-tolerant, attack-resilient and reliable systems. A detailed discussion of this is beyond the scope of this chapter. These innovations are often opaque to the user. As such, we provide a high-level overview of approaches to reliability including ensuring reliability by design through monitoring, redundancy and disaster recovery, and the evaluation of performance and quality of service (QoS).

A major focus of computer science research is reliability by design so that no one point of failure can result in the failure of the entire system. There are a wide variety of causes of unplanned cloud outages including infrastructure or software failures, planning mistakes, human error, or external attacks (Endo et al. 2017). Three main strategies are employed to counter such failures namely, monitoring, redundancy, and disaster recovery. In the terminology of trust, two could be classified as trust-building mechanisms (monitoring and redundancy) while the third, disaster recovery, could be classified as a trust repair mechanism. A wide variety of general purpose and vendor-specific monitoring tools are used in cloud computing. From the user perspective, these are primarily used for accounting and billing, security and privacy assurance, and SLA management, while for the cloud service provider they may be used for other reliability functions, for example fault management (Fatema et al. 2014). As mentioned earlier, gray failures may not be detectable by extant monitoring systems that focus on singular failure detection. To mitigate the risk of such failures, Huang et al. (2017) suggest that cloud service providers must move to multi-dimensional cloud health monitoring. While accepting monitoring all applications and workloads in hyperscale multi-tenant systems is not feasible, they propose a number of techniques to close the observation gap including approximating application views, aggregating observations from multiple components to infer the likelihood of a gray failure in an isolated component, as well as temporal analysis (Huang et al. 2017). As noted briefly in Chap. 1, monitoring data can be used more widely in the context of building knowledge-based trust. Emeakaroha et al. (2016) have proposed a system and show through experimental studies with business decision-makers that such monitoring systems can be used to build trust through communication strategies such as trust labels (Emeakaroha et al. 2016; van der Werff et al. 2018).

Cloud failures can be caused by issues that occur at different levels in the cloud stack e.g. at the data, application, and/or system level (Huang et al. 2017). Given organisational and consumer concerns about data and availability of data in the event of a failure, it is unsurprising that in addition to general system redundancy, data redundancy is a primary concern of cloud service providers. Data replication and erasure coding are commonly used data redundancy techniques in cloud computing (Nachiappan et al. 2017). With simple data replication, data is replicated in at least two locations on distributed cloud storage systems so that in the event of storage failure, it is just served from the replicated copy (Plank 2013). As such, data loss only occurs if data corrupted on all storage targets the replicated copies (Rajaasekharan 2014). As simple data replication carries a significant resource overhead in terms of storage, network and associated energy consumption, hyperscale cloud service providers, such as Facebook and Microsoft, use more advanced erasure coding, such as K out of N codes, to detect and correct errors in cloud storage, and provide a less resource intensive means to reconstruct data from parity data (Nachiappan et al. 2017; Rajaasekharan 2014).

Disasters differ in terms of scale and impact (although this is subjective), and are typically unpredicted events that occur relatively rarely over the lifetime of a given system. A full cloud service outage occurs more frequently than one might imagine but due to the disaster recovery systems in place, the recovery time is extremely fast. Disasters can result from natural, human, or technological causes, or a combination of two or more of these (Singh et al. 2016). To mitigate the impact of natural disasters or large-scale malicious physical attacks, cloud service providers, like many IT organisations, use distributed backups, online and offline, in geographic locations that are located sufficiently distant to avoid a homogenous natural event (Pokharel et al. 2010). Maintaining two infrastructures is extremely costly. However, cloud outages can also result from relatively small-scale localised natural causes, for example lightning strikes are a significant threat to both primary and uninterruptible power supply (Li et al. 2013). Human causes include human error or malicious attacks from insiders or external third parties. The latter is largely a security issue while the former is a training and behavioural one. Li et al. (2013) document a wide range of public cloud outages resulting from human error including vehicle accidents, power shutdowns, and inputting commands in error. As discussed earlier in this section, application and system level failures can be technological causes of full service outage. In these instances, for application failures, the key requirement is business continuity through redundancy and rollback. It should be noted that a number of middleware approaches have been applied to address application-level reliability via application-independent failure detection, checkpoint and rollback and recovery (e.g. Hormati, et al. 2014), optimal replica placement (e.g. An et al. 2014), stop and copy VM migration (Sampaio and Barbosa 2018), and entity reputation management (Abawajy 2011). For system level failures, the primary focus is minimising recovery time (Singh et al. 2016). It is important to note that while these causes are isolated, they may be cascading, natural causes can result in unanticipated technological failures, which in turn may be exacerbated by human errors, and so forth.

As discussed in Chap. 2, the SLA details the level of service to be provided, often in the form of specific QoS metrics (Ghazizadeh and Cusack 2018). Obviously, in the context of trust, there is a close relationship between SLA metrics and monitoring, and unsurprisingly this is a major focus of both cloud monitoring systems (see Fatema et al. 2014) and trustworthy cloud computing research. This research primarily focuses on the decomposition of SLA parameters in to low-level system performance metrics, mapping these in to KPIs, and then ultimately aggregating these KPIs in to some form of aggregated quality indicator that can be used to mitigate transactional risk (Sun et al. 2012). A wide range of techniques are used to measure and predict cloud service performance (and indeed SLA violation). Typical metrics include availability, bandwidth, cost (including energy), CPU cycle, service duration, memory, request arrival rate, space/storage. Upgrade request frequency as well as other more specific performance metrics (throughput, response time, execution time etc.) are also present, although the importance of these will vary by cloud service (Faniyi and Bahsoon 2015). Cloud service providers may also include metrics that specifically acknowledge the risk of failure e.g. the maximum fraction of SLA violations allowed or penalty rates (Faniyi and Bahsoon 2015). Notably, security is an attribute metric that is extremely difficult to measure and is typically based on a qualitative evaluation of cloud service provider policies and system features (Shaikh and Sasikumar 2015). Once such metrics have been extracted from the system, they can be shared with consumers to build trust or select cloud service providers. An example of the former is the cloud trust label mentioned earlier (Emeakaroha et al. 2016; van der Werff et al. 2018). Regarding the latter, Garg, et al. (2013) propose a Service Measurement Index Cloud (SMICloud) framework for assisting consumers to identify the most suitable cloud service provider to contract with. The SMICloud reviews Quality of Service (QoS) requirements and ranks services based on previous user experiences and performance of services based on KPIs such as those previously mentioned. As a final note on cloud performance metrics, the determination of the intervals for this data is an essential and somewhat open challenge. This includes the monitoring intervals between the collection of low-level metrics and the intervals between the aggregate KPIs or high-level quality indicators (Sun et al. 2012). A balance between intrusiveness and utility is required to avoid adverse impacts on system performance while ensuring the availability of sufficiently time-sensitive data to assure accurate SLA measurement (Sun et al. 2012).

7.4   BUSINESS INTEGRITY

As discussed in Chap. 1, the trust literature views integrity generally as one party's perception that another party will adhere to a set of acceptable principles, act honestly, and fulfil their promises (Mayer et al. 1995; McKnight et al. 2011). This is consistent with the principles laid out by Microsoft in Mundie et al. (2002), namely that a vendor, in this case a cloud service provider, will behave in a responsive and responsible manner. While Mundie et al. (2002) exemplify this behaviour in terms of responsiveness to problems that may arise, others expand this, in a technological context, to mean that both the service and vendor behave predictably to the extent which it is possible to anticipate the system and the service provider's behaviour accurately (van der Werff et al. 2018). In one sense, it is no surprise that computer scientists have found it difficult to distinguish reliability, as an attribute, from integrity.

In computer science literature, integrity is more commonly found as an attribute of data and underlying systems rather than the service as a whole or the vendor. This is not to say that computer science researchers have not explored technological innovations in this regard. In addition to attempts to communicate performance metrics and service measurement mechanisms similar to those outlined in Sect. 7.3 above, some researchers have focussed on more holistic evaluations of cloud services and service providers. As referenced briefly in Chap. 1, feedback systems and reputation management systems are two approaches explored in research to build trust. For example, Baranwal and Vidyarthi (2014) propose a Service Measurement Index (SMI) comprising two sets of metrics—application-dependent metrics and user-dependent metrics. Notably, in the context of Mundie et al. (2002), they include customer support as an application-dependent metric. Unlike the SLA-focused measurements discussed earlier, SMI includes reputation metrics based on feedback from users, user experience and certification of compliance with industry best practice and regulations. In a similar vein, Machhi and Jethava (2016) present a trust management framework that measures service provider trustworthiness based on feedback, aging factor, and other parameters, while eliminating or otherwise discounting unreliable feedback. Indeed a number of works have sought to combine SLA metrics with feedback systems as a means of communicating trust in the service and vendor (see, for example, Nguyen et al. 2010; Habib et al. 2011; Yau and Yin 2011; Garg et al. 2013; Noor, et al. 2015; Tang et al. 2017).

While these researchers have sought to explore integrity as a quantifiable attribute of a service, business integrity is typically either conflated as competence (see for example Chakraborty and Roy 2012), or as a function of information assurance practices and qualitative audits such as certification (Chakraborty et al. 2010).

7.5   Conclusion

This chapter presented a discussion on trustworthy computing from three perspectives—security and privacy, reliability, and business integrity. Computer science research has typically sought to focus on trust as an objective attribute of systems, and on occasion cloud service providers, that can be ultimately measured, compared and benchmarked. One might argue that it is a narrow view of trust that misses the more nuanced aspects of the psychological underpinnings of trust. This may go some way to explaining why trust remains a significant barrier to cloud computing adoption. As a starting point, researchers might consider using the taxonomy of trustworthy computing laid out by Microsoft in Mundie et al. (2002), i.e. goals, means and execution, to identify gaps in the literature and state of the art, and guide future avenues for research. As we move towards the Internet of Things, and greater use of advanced autonomous technologies, such as self-learning, self-management, and artificial intelligence, a more inter- and multi-disciplinary approach is needed to ensure that all stakeholders benefit fully and fairly from these transformative technologies.

REFERENCES

Abawajy, J. (2011). *Establishing Trust in Hybrid Cloud Computing Environments.* 2011IEEE 10th International Conference on Trust, Security and Privacy in Computing and Communications (pp. 118–125). IEEE.

Abdalla, A. K. A., & Pathan, A. S. K. (2014). On Protecting Data Storage in Mobile Cloud Computing Paradigm. *IETE Technical Review, 31*(1), 82–91.

Acar, A., Aksu, H., Uluagac, A. S., & Conti, M. (2018). A Survey on Homomorphic Encryption Schemes: Theory and Implementation. *ACM Computing Surveys (CSUR), 51*(4), 1–35.

Adams, M., Bearly, S., Bills, D., Foy, S., Li, M., Rains, T., Ray, M., Rogers, D., Simorjay, F., Suthers, S., & Wescott, J. (2014). *An Introduction to Designing Reliable Cloud Services.* Redmond, WA: Microsoft Corporation.

Ali, M., Dhamotharan, R., Khan, E., Khan, S. U., Vasilakos, A. V., Li, K., & Zomaya, A. Y. (2015). SeDaSC: Secure Data Sharing in Clouds. *IEEE Systems Journal, 11*(2), 395–404.

Alqahtani, H. S., & Kouadri-Mostefaou, G. (2014). *Multi-clouds Mobile Computing for the Secure Storage of Data.* 2014 IEEE/ACM 7th International Conference on Utility and Cloud Computing (pp. 495–496). IEEE.

An, K., Shekhar, S., Caglar, F., Gokhale, A., & Sastry, S. (2014). A Cloud Middleware for Assuring Performance and High Availability of Soft Real-Time Applications. *Journal of Systems Architecture, 60*(9), 757–769.

Baranwal, G., & Vidyarthi, D. P. (2014). *A Framework for Selection of Best Cloud Service Provider Using Ranked Voting Method.* 2014 IEEE International Advance Computing Conference (IACC) (pp. 831–837). IEEE.

Bucur, V., Dehelean, C., & Miclea, L. (2018). *Object Storage in the Cloud and Multi-cloud: State of the Art and the Research Challenges.* 2018 IEEE International Conference on Automation, Quality and Testing, Robotics (AQTR) (pp. 1–6). IEEE.

Chakraborty, R., Ramireddy, S., Raghu, T. S., & Rao, H. R. (2010). The Information Assurance Practices of Cloud Computing Vendors. *IT Professional, 12*(4), 29–37.

Chakraborty, S., & Roy, K. (2012). *An SLA-based Framework for Estimating Trustworthiness of a Cloud.* 2012 IEEE 11th International Conference on Trust, Security and Privacy in Computing and Communications (pp. 937–942). IEEE.

Dummer, G. W. A., Winton, R., & Tooley, M. (1997). *An Elementary Guide to Reliability*. Elsevier.

Emeakaroha, V. C., Fatema, K., van der Werff, L., Healy, P., Lynn, T., & Morrison, J. P. (2016). A Trust Label System for Communicating Trust in Cloud Services. *IEEE Transactions on Services Computing, 10*(5), 689–700.

Endo, P. T., Santos, G. L., Rosendo, D., Gomes, D. M., Moreira, A., Kelner, J., Sadok, D., Gonclaves, G. E., & Mahloo, M. (2017). Minimizing and Managing Cloud Failures. *Computer, 50*(11), 86–90.

European Commission. (2020). Shaping Europe's Digital Future. Retrieved June 6, 2020, from https://ec.europa.eu/info/sites/info/files/communication-shaping-europes-digital-future-feb2020_en_4.pdf

Faniyi, F., & Bahsoon, R. (2015). A Systematic Review of Service Level Management in the Cloud. *ACM Computing Surveys (CSUR), 48*(3), 1–27.

Fatema, K., Emeakaroha, V. C., Healy, P. D., Morrison, J. P., & Lynn, T. (2014). A Survey of Cloud Monitoring Tools: Taxonomy, Capabilities and Objectives. *Journal of Parallel and Distributed Computing, 74*(10), 2918–2933.

Garg, S. K., Versteeg, S., & Buyya, R. (2013). A Framework for Ranking of Cloud Computing Services. *Future Generation Computer Systems, 29*(4), 1012–1023.

Gates, B. (2002). Trustworthy Computing. Retrieved June 6, 2020, from https://www.wired.com/2002/01/bill-gates-trustworthy-computing/

Ghazizadeh, E., & Cusack, B. (2018). Evaluation Theory for Characteristics of Cloud Identity Trust Framework. In *Cloud Computing-Technology and Practices*. IntechOpen.

Habib, S. M., Ries, S., & Muhlhauser, M. (2011). *Towards a Trust Management System for Cloud Computing*. 2011 IEEE 10th International Conference on Trust, Security and Privacy in Computing and Communications (pp. 933–939). IEEE.

Hayden, M. V. (2000). *National Information Systems Security Glossary*. National Security Telecommunications and Information Systems Security Committee.

Hormati, M., Khendek, F., & Toeroe, M. (2014). *Towards an Evaluation Framework for Availability Solutions in the Cloud*. 2014 IEEE International Symposium on Software Reliability Engineering Workshops (pp. 43–46). IEEE.

Huang, P., Guo, C., Zhou, L., Lorch, J. R., Dang, Y., Chintalapati, M., & Yao, R. (2017). *Gray Failure: The Achilles' Heel of Cloud-Scale Systems*. Proceedings of the 16th Workshop on Hot Topics in Operating Systems (pp. 150–155).

Khan, A. N., Kiah, M. M., Ali, M., Madani, S. A., & Shamshirband, S. (2014). BSS: Block-based Sharing Scheme for Secure Data Storage Services in Mobile Cloud Environment. *The Journal of Supercomputing, 70*(2), 946–976.

Li, Z., Liang, M., O'Brien, L., & Zhang, H. (2013). The Cloud's Cloudy Moment: A Systematic Survey of Public Cloud Service Outage. *International Journal of Cloud Computing and Services Science, 2*(5), 1–15.

Liu, J. K., Au, M. H., Susilo, W., Liang, K., Lu, R., & Srinivasan, B. (2015). Secure Sharing and Searching for Real-Time Video Data in Mobile Cloud. *IEEE Network, 29*(2), 46–50.

Louk, M., & Lim, H. (2015). *Homomorphic Encryption in Mobile Multi Cloud Computing.* 2015 International Conference on Information Networking (ICOIN) (pp. 493–497). IEEE.

Machhi, S., & Jethava, G. B. (2016). *Feedback Based Trust Management for Cloud Environment.* Proceedings of the Second International Conference on Information and Communication Technology for Competitive Strategies (pp. 1–5).

Mayer, R. C., Davis, J. H., & Schoorman, F. D. (1995). An Integrative Model of Organizational Trust. *Academy of Management Review, 20*(3), 709–734.

McKnight, D. H., Carter, M., Thatcher, J. B., & Clay, P. F. (2011). Trust in a Specific Technology: An Investigation of Its Components and Measures. *ACM Transactions on Management Information Systems (TMIS), 2*(2), 12.

Mollah, M. B., Azad, M. A. K., & Vasilakos, A. (2017). Secure Data Sharing and Searching at the Edge of Cloud-Assisted Internet of Things. *IEEE Cloud Computing, 4*(1), 34–42.

Moore, C., O'Neill, M., O'Sullivan, E., Doröz, Y., & Sunar, B. (2014). *Practical Homomorphic Encryption: A Survey.* Circuits and Systems (ISCAS), 2014 IEEE International Symposium on (pp. 2792–2795). IEEE Computer Society. https://doi.org/10.1109/ISCAS.2014.6865753

Mundie, C., de Vries, P., Haynes, P., & Corwine, M. (2002). *Trustworthy Computing.* Technical Report, 10.

Nachiappan, R., Javadi, B., Calheiros, R. N., & Matawie, K. M. (2017). Cloud Storage Reliability for Big Data Applications: A State of the Art Survey. *Journal of Network and Computer Applications, 97*, 35–47.

Nguyen, H. T., Zhao, W., & Yang, J. (2010). *A Trust and Reputation Model Based on Bayesian Network for Web Services.* 2010 IEEE International Conference on Web Services (pp. 251–258). IEEE.

Noor, T. H., Sheng, Q. Z., Yao, L., Dustdar, S., & Ngu, A. H. (2015). CloudArmor: Supporting Reputation-Based Trust Management for Cloud Services. *IEEE Transactions on Parallel and Distributed Systems, 27*(2), 367–380.

Plank, J. S. (2013). Erasure Codes for Storage Systems: a Brief Primer. *Usenix Magazine, 38* (6), 44–50.

Pokharel, M., Lee, S., & Park, J. S. (2010). *Disaster Recovery for System Architecture Using Cloud Computing.* 2010 10th IEEE/IPSJ International Symposium on Applications and the Internet (pp. 304–307). IEEE.

Rajaasekharan, A. (2014). Data Reliability in Highly Fault-Tolerant Cloud Systems. Seagate. Retrieved June 6, 2020, from https://www.seagate.com/files/www-content/_shared/_masters/category-info/data-reliability-fault-tolerant-cloud-pv0031-1-1410-us.pdf

Sampaio, A. M., & Barbosa, J. G. (2018). A Comparative Cost Analysis of Fault-Tolerance Mechanisms for Availability on the Cloud. *Sustainable Computing: Informatics and Systems, 19*, 315–323.

Shaikh, R., & Sasikumar, M. (2015). Trust Model for Measuring Security Strength of Cloud Computing Service. *Procedia Computer Science, 45*, 380–389.

Singh, S., Jeong, Y. S., & Park, J. H. (2016). A Survey on Cloud Computing Security: Issues, Threats, and Solutions. *Journal of Network and Computer Applications, 75*, 200–222.

Sookhak, M., Akhunzada, A., Gani, A., Khurram Khan, M., & Anuar, N. B. (2014). Towards Dynamic Remote Data Auditing in Computational Clouds. *The Scientific World Journal*, 2014.

Sun, L., Singh, J., & Hussain, O. K. (2012). *Service Level Agreement (SLA) Assurance for Cloud Services: A Survey from a Transactional Risk Perspective.* Proceedings of the 10th International Conference on Advances in Mobile Computing & Multimedia (pp. 263–266).

Tan, J., Gandhi, R., & Narasimhan, P. (2014). *STOVE: Strict, Observable, Verifiable Data and Execution Models for Untrusted Applications.* 2014 IEEE 6th International Conference on Cloud Computing Technology and Science (pp. 644–649). IEEE.

Tang, M., Dai, X., Liu, J., & Chen, J. (2017). Towards a Trust Evaluation Middleware for Cloud Service Selection. *Future Generation Computer Systems, 74*, 302–312.

Tebaa, M., & Hajji, S. Â. E. (2014). Secure Cloud Computing Through Homomorphic Encryption. Preprint arXiv:1409.0829. [Online]. Available: https://arxiv.org/abs/1409.0829.

Tian, H., Chen, Y., Chang, C. C., Jiang, H., Huang, Y., Chen, Y., & Liu, J. (2015). Dynamic-Hash-Table Based Public Auditing for Secure Cloud Storage. *IEEE Transactions on Services Computing, 10*(5), 701–714.

van der Werff, L., Real, C., & Lynn, T. (2018). Individual Trust and the Internet. In R. H. Searle, A. M. I. Nienaber, & S. B. Sitkin (Eds.), *The Routledge Companion to Trust* (pp. 391–407). Abingdon: Routledge.

Wang, H., Wu, S., Chen, M., & Wang, W. (2014). Security Protection between Users and the Mobile Media Cloud. *IEEE Communications Magazine, 52*(3), 73–79.

Yang, Y., Liu, J. K., Liang, K., Choo, K. K. R., & Zhou, J. (2015). *Extended Proxy-Assisted Approach: Achieving Revocable Fine-Grained Encryption of Cloud Data.* European Symposium on Research in Computer Security (pp. 146–166). Cham: Springer.

Yau, S. S., & Yin, Y. (2011). *QoS-based Service Ranking and Selection for Service-Based Systems.* 2011 IEEE International Conference on Services Computing (pp. 56–63). IEEE.

Yu, Y., Li, Y., Au, M. H., Susilo, W., Choo, K. K. R., & Zhang, X. (2016). *Public Cloud Data Auditing with Practical Key Update and Zero Knowledge Privacy.* Australasian Conference on Information Security and Privacy (pp. 389–405). Cham: Springer.

Yu, Y., Mu, Y., Ni, J., Deng, J., & Huang, K. (2015). *Identity Privacy-Preserving Public Auditing with Dynamic Group for Secure Mobile Cloud Storage.* International Conference on Network and System Security (pp. 28–40). Cham: Springer.

INDEX